A Guide to Bird Behavior

Volume I

A Guide to Bird Behavior

Volume I

Donald W. Stokes

Illustrated by the author

with 25 original illustrations by J. Fenwick Lansdowne

Little, Brown and Company Boston Toronto

Fourth Printing

B

Library of Congress Cataloging in Publication Data
Stokes, Donald W.
A guide to bird behavior .

Bibliography: p.
1. Birds — Behavior. 2. Birds — North America — Behavior. I. Lansdowne, James Fenwick. II. Title.
QL698.3.S76 598.2'5 79-17864
ISBN 0-316-81722-8

MV
Designed by Susan Windheim

Published simultaneously in Canada
by Little, Brown & Company (Canada) Limited

PRINTED IN THE UNITED STATES OF AMERICA

Acknowledgments

THE FIRST AND MOST IMPORTANT ACKNOWLEDGMENT MUST GO TO the many professional and amateur ornithologists who took the time to study our common birds. This book could never have been done without their initial work, and their names and studies are fully listed in the bibliography. Next I would like to thank John T. Emlen, Jr., and Millicent S. Ficken for their careful readings of the manuscript and their many expert suggestions for its improvement. I also want to thank my editor at Little, Brown and Company, William D. Phillips, for helping me to make the final book a realization of my initial design. And last, I would like to thank Steve Howard for his enthusiasm for this project when we first went out to the duck pond to look at Mallard behavior.

Contents

A Guide to Bird Behavior

Behavior-Watching

I HAVE ALWAYS ENJOYED THE CONTINUAL TREASURE HUNT OF BIRD-watching — going out in the early morning and searching grasses, shrubs and trees for the least movement, quickly spotting the familiar birds, and puzzling over new spring arrivals or fall migrants. But at the same time, just making lists of birds I'd seen left me feeling somewhat unsatisfied — I sensed I wasn't getting to the heart of the matter, was somehow passing up an opportunity or neglecting a resource. This feeling was always strongest in winter, when the few species that remained were common and easy to spot, so that the element of surprise and the treasure hunt that I had enjoyed in other seasons were no longer present. What could I do with a bird once I knew its name? Was there some other aspect of its life that I could discover in winter? Or was I dependent on new spring arrivals to restore my interest and keep me involved? These questions puzzled me, and I had no easy answers.

Then, early one winter a few years ago, I found an article by Konrad Lorenz on Mallards. It was filled with illustrations of the birds in strange poses, labeled with captions like "Grunt-whistle," "Inciting," and "Nod-swimming." I had learned to identify Mallards long ago and felt that I knew them well, but I had never seen them do anything like the poses that were illustrated in the article. Were these things common and easy to observe? And if so, how could I have missed them all these years? I soon went to a local duck pond to get the answer.

I arrived with a naturalist friend, and we began looking over the ducks. Within a few minutes we began to see some of the very things pictured in the article — males arching their necks and giving a whistlelike call, females jerking their heads to one side as they followed males, pairs of birds alternately bobbing their heads up and down as they faced each other, and females darting around males with their heads held close to the water. These displays were happening all about the pond, and we could barely keep our eyes on one without being distracted by another.

We were amazed by what we saw and our excitement was obvious, for other people approached us wanting to know what new species we had seen. The experience of being thrilled by the behavior of a common bird was so new and unexpected that we had trouble explaining it to those present, especially when they believed, as we had before, that only rare species could be this exciting. Indeed the Mallard had become a new bird for me, a bird I had never really seen before, and I was already looking forward to spending more time searching for other elements of its behavior. The surprise and treasure hunt had been restored to my winter bird-watching. I no longer felt empty, but full of questions and wonder at the things I had seen.

After having my view of Mallards transformed, I was curious to find if other common birds were equally intriguing in their life habits. I looked for more articles in ornithological journals and began to discover many on the behavior of our common species, each piece full of fascinating observations.

I was surprised to find that very few people were aware of social behavior in birds. Even those who had been experienced bird-watchers for decades knew nothing of the Mallard displays. This lack of public knowledge, my experience with the Mallards, and my discovery of the research all led me to the writing of this field guide. It is the result of my desire to share with others the endless surprise and discovery produced by behavior-watching.

The behavior of birds can be divided into two broad categories: maintenance behavior and social behavior. The former includes all actions a bird does to maintain itself, such as preening, feeding, bathing, etc.; the latter includes all interactions between birds, such as courtship, territoriality, breeding, flocking, etc. This guide deals primarily with social behavior, for this is one of the most exciting aspects of birds, and one that has often been neglected. In social behavior one individual is coordinating its life and actions with those of another, and this always involves communication of some kind, either in fixed sounds or gestures, or in general movement patterns. This struggle to connect separate lives is common to all animal life, and I find that through watching it in birds, I have become more aware of its importance in the lives of all animals.

This application of behavior-watching is one of its deepest rewards, but at the same time it is one of its most common pitfalls; for although behavior-watching can produce insights into human behavior, applying human motives to birds can be very limiting. When most of us try to supply motives for the behavior of animals, we invariably use motives that we would have in the same situation. "The bird scolded me." "The mother bird was teaching its young how to fly." "He sang happily from his perch." All of these statements are anthropomorphic: they are assumptions about the motives of birds based on human values. This tendency to explain the actions of birds in human terms greatly limits our ability to learn new things about the avian world. It is like searching the bottom of a pond for plants and animals and never getting past our own image on the water's surface.

What a bird does and why it does it are two different things. One is observation and therefore mostly fact, the other is assumption and so mostly speculation. The problem is that we have a strong tendency to mix the two and therefore confuse facts with assumptions. In your own observations, keep descriptions of behavior and conclusions about the meaning of the behavior separate. If you do this, then your observations will be of greater value to yourself, to other behavior-watchers, and even to the scientific community.

When I go out to behavior-watch I usually take along a pair

of binoculars, an identification guide, and a small notebook for writing down what I see and hear. I start looking and listening as soon as I get out the door, whether in the city or the country, for fascinating behavior can happen anywhere. I listen for all bird sounds and keep my eyes open for all interactions — both of these are the main indicators of social behavior.

If I see an interaction I try to answer the following questions about it: What species is/are interacting? What happens before, during and after the interaction? How long does it last? Is it repeated? Is it conspicuous or hard to see? How many birds are involved? Are all birds in the interaction behaving alike?

Often when birds are interacting, they make sounds or assume unusual positions with their bodies and/or feathers. These auditory or visual displays are part of the birds' language. They are among the most exciting and important aspects of bird behavior, and learning to recognize them is essential to understanding the interactions of birds. When I see what I believe to be a display, I try to answer these questions: What species is displaying? How many different displays is it using? How will I distinguish between the different displays of one bird? How long does the display last? What does the bird do before, during and after the display? Do other birds or other animals seem to respond to the display? Does the display include both gestures and sounds?

In all cases it is also helpful to make these further observations: the date, the time of day, where the behavior took place (i.e., in the territory, near the nest, near a mate, etc.), and whether the bird is male or female, young or adult.

As well as observing specific behavior, I always take a general overall impression of the bird I am watching. Is it alone, in pairs, small flocks, or large flocks? Is it conspicuous or secretive? Is its display repertoire large or small? Over the months a simple record of these features will add up to a great deal of information about the behavior of each bird.

The social behavior of most birds is periodic, so it is often productive to be patient and wait through periods of inactivity.

It is also important to be aware of how your own presence is affecting the birds you are watching. In most cases you will want to stay far enough away so that you don't disturb their natural patterns, but at other times you may want to be closer and see how the birds respond. The majority of social behavior occurs from sunrise until about 11:00 A.M. and then again from about 3:00 P.M. until sunset. In midday birds are generally quiet and hard to locate.

These suggestions are only guidelines for your own explorations. Some of you will find that just watching bird behavior and using this guide to help interpret what you see will be enough. Others may want to make careful records of their observations and not only discover all they can about the birds in this guide, but continue learning about other species as well.

The most important thing to remember in all of this is that the discovery of bird behavior is open to all. The general public believes that all of the common aspects of nature are already well known. Nothing could be further from the truth in the field of bird behavior. This was one of the biggest surprises to me when I started writing this book. There were many common birds I would have loved to include, but they simply had never been studied. Even with the birds that have been well studied, such as the Song Sparrow, Red-Winged Blackbird and Mallard, there are still far more mysteries than answers concerning their behavior. In fact, in most cases there has not been enough observation to know for sure what is individual behavior and what is the general behavior of the species. Because of this, what *you* see birds do is very important. Do not discount it; remember it and record it. One of the main purposes of this guide is to encourage everybody to participate in helping to discover the behavior of our common birds. If just a fraction of the energy that now goes into bird-watching were to go into behavior-watching, within a very short time our knowledge about the behavior of our common birds would be greatly increased.

How to Use This Book

TO DISCOVER AND INTERPRET A BIRD'S BEHAVIOR YOU NEED THREE types of information: the general timing of the bird's life stages, knowledge of its displays used to communicate with other birds, and the details of its major behavior patterns. To answer these needs, the information on each bird in this guide is organized into three sections: a behavior calendar, a display guide, and behavior descriptions.

The behavior descriptions form the main body of information. They describe in detail six major areas of each bird's behavior: territory, courtship, nest-building, breeding, plumage, and seasonal movement. An additional section on social behavior is included for those species that have significant group behavior outside of the above categories.

Preceding each bird's behavior descriptions are two concise references — the behavior calendar and the display guide. The behavior calendar lists the major areas of a bird's behavior next to the appropriate months. The display guide provides pictures and descriptions for the field identification of the main displays used by each species. Both of these short guides will refer you to sections of the behavior descriptions for further information on what you have seen.

BEHAVIOR CALENDAR

The behavior calendar is only an approximation of the timing of the bird's life cycle and is meant to give the observer a rough idea of when the behaviors described are most common. Clearly, timing varies widely within North America. Because of this, the behavior calendar has been calculated for the middle latitudes of the continent (around forty degrees latitude, or, roughly, a line drawn from Philadelphia, through Indianapolis and Denver, to slightly above San Francisco).

It has been found that breeding times vary ten to fifteen days for each change of five degrees latitude. Therefore, with the aid of the map below, you can quickly become accustomed to adjusting the behavior calendar for whatever area of the continent you are in. As you continue to observe behavior patterns, you may want to record a more accurate estimate of their timing for your particular area.

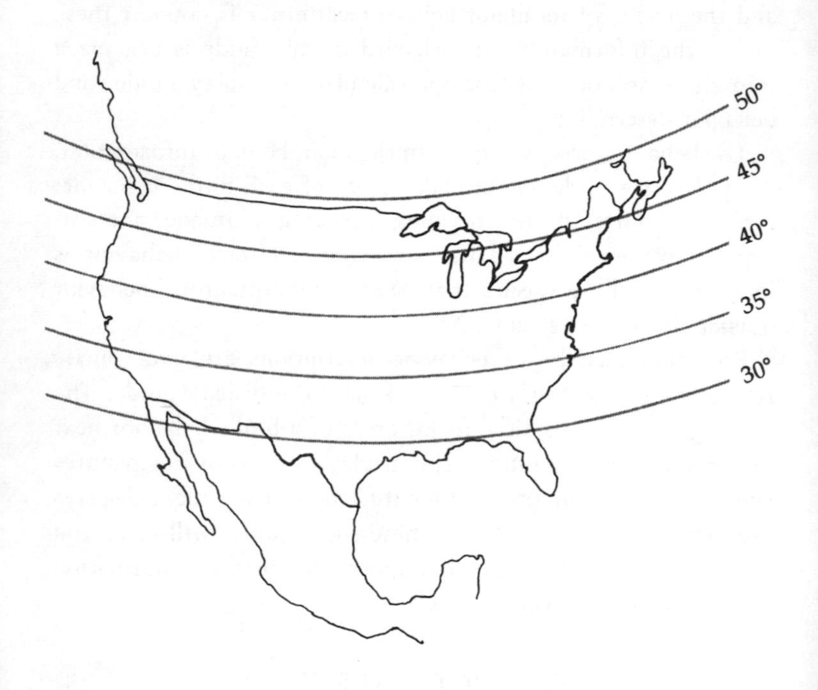

DISPLAY GUIDE

The display guide lists the major displays used by each species.

A display is a gesture or sound that a bird makes and that, when used in certain situations, affects the behavior of other animals near the bird. More loosely, it is a signal birds use to communicate with each other.

Bird displays are generally divided into two groups: those that are heard (auditory displays) and those that are seen (visual displays). Visual displays have been illustrated in those cases where a picture is particularly helpful, and the auditory displays have been represented with verbal approximations whenever feasible. When a call usually accompanies a visual display, it is listed below it. When no call is listed, the display is usually accomplished in silence.

The verbal approximations of the calls have been punctuated in four different ways to show the timing of the sounds:

tseetseetseetsee — spoken continuously with no pause
tsee tsee tsee — slight pause
tseet, tseet, — moderate pause
tseet. tseet. — substantial pause

The descriptions of the auditory displays are not sufficiently detailed to enable you to distinguish the calls of one species from those of another. But this is not as important for behavior-watching as is being able to distinguish the different calls of the same species. For this purpose the display guide will be extremely helpful. After most of the displays, you will see a reference to one or more sections of the behavior descriptions, such as, "*See* Territory, Courtship." By reading these sections you will understand more fully the use and context of the display. Displays that are used throughout the year may have no reference to a specific section of the behavior descriptions.

The seasons in which each display is commonly given are listed with the display and represented by symbols: Sp — spring, Su — summer, F — fall, and W — winter.

BEHAVIOR DESCRIPTIONS

Territory

Territory has been defined as "any defended area." It is important to remember that territory is not the same for all birds; it is extremely varied. Territory is categorized into various types, depending on what occurs in it. Some territories include only the nest and the two or three feet of space around it; other territories include a larger area for mating and/or feeding.

Courtship

In this guide *courtship* refers to pair formation, pair maintenance, and mating. *Pair formation* includes the initial contacts between a male and female of the same species, both in breeding condition. Often, after a pair of birds are committed to each other, they continue to perform displays when they meet, or they may have extended periods of mate-feeding. This is *pair maintenance*, or the sustaining of the relationship and commitment between the two birds.

Mating is used here to refer to copulation as well as the displays that come before and after it. Copulation is similar in all common birds. The female usually tilts forward and raises her tail feathers while the male steps onto her back and lowers his tail feathers to make contact with her. In most birds, copulation lasts for only a few seconds, but may be repeated many times. The displays surrounding copulation are often some of the most elaborate in the birds' repertoire.

Nest-Building

This section offers a guide to locating the nest of each species. To those of you who have not done it before, this may seem an impossible task, but with the help of the behavioral clues to the nest location, you will find it easier than you expect. With some species, finding the nest is essential for seeing the most interesting elements of the birds' behavior.

Breeding

This is the period from the beginning of egg-laying until the time when the young are independent. It is divided into three phases: egg-laying and incubation, the nestling phase, and the fledgling phase. The nestling phase is the time from when the young hatch until they leave the nest. The fledgling phase lasts from the time they leave the nest until they are no longer dependent upon the parents for food.

Plumage

Though plumage is not an aspect of the birds' social behavior, it is helpful when watching behavior to know when the molting periods occur and also to know the various changes in the appearance of the bird through the year. This section also gives clues that help in distinguishing the sexes by using physical appearance as well as behavior.

Seasonal Movement

This section describes the large-scale movements of birds that regularly occur each year. The largest seasonal movement of most birds is migration, but nonmigratory as well as migratory birds often shift slightly from one environment or area to another, and I have described these lesser movements where significant.

For only a few species in this guide is migration a clear-cut phenomenon — birds flying south in the fall and north in the spring. For the majority of birds in this guide, migration patterns vary within a species, some members migrating long distances, others migrating short distances, and still others remaining as year-round residents. Some birds even migrate one year and stay put the next. The problems of studying migration are immense with the result that there is still very little known about this aspect of our common birds.

Social Behavior

Social behavior normally refers to all interactions between birds, but here it has the restricted meaning of all interactions between nonbreeding birds of the same species, or those outside of their breeding period (in other words, all interactions that are not already discussed in the previous sections). This type of behavior usually occurs in fall and winter, and involves flocks that coordinate their actions in some way.

Feeder Behavior

Home bird feeders are an ideal place to observe and learn about bird behavior. A Feeder Behavior section has been provided for those birds that regularly visit feeders. In each are described the bird's favorite foods, its general behavior at the feeder, its most common displays given in feeder situations, and other behavior that you may see in the vicinity of the feeder. Reading the Feeder Behavior sections of this guide is probably the best way to begin your study of bird behavior, especially if you are starting in fall and winter, when the most birds are attracted to feeders.

This guide is designed to help you interpret what you see in the field. For example, say you have just walked by a swamp in March and seen a number of Red-Winged Blackbirds perched on the tops of the cattails. The birds called continually and there were frequent chases.

To understand what was happening, first turn to the behavior calendar of the Red-Winged Blackbird and see what behavior occurs in March. You will find Territory and Seasonal Movement shaded in for that month. Now turn to the behavior descriptions for the Red-Winged Blackbird and read the sections on territory and on seasonal movement. Here you are most likely to find a detailed description of the behaviors that you saw.

If you observe the birds closely enough to see unusual postures (visual displays) or to hear distinct calls (auditory displays), turn to the display guide, and it will help you to identify them. Generally, following a display there will be a reference to the section of the behavior descriptions where there is further information on the meaning of the display.

Besides using the guide in the field, you may want to consult it before going out, to familiarize yourself with the behavior of several birds that you know you will be able to observe. In this case, read at least the introduction to each bird. If you have more time, read the behavior descriptions that are appropriate for that month, and get to know the distinctions between the various displays.

Not all aspects of a bird's behavior can be learned with one reading or one session of behavior-watching. More likely, you will alternate between observing in the field and reading in the guide, each time learning a little more about the bird's behavior. A glossary is provided at the end of the book to help explain any unfamiliar terms.

If the birds in this guide are not already familiar to you, the drawings at the beginning of each chapter will help you to identify them. For expert field identification of these and other birds, the best book to have is *Birds of North America* by Chandler S. Robbins, Bertel Bruun, and Herbert S. Zim.

Canada Goose / *Branta canadensis*

IF YOU ARE NEAR A PAIR OF CANADA GEESE DURING THE BREEDING season, you are bound to witness a conspicuous Greeting Ceremony. This is a series of visual and auditory displays given by a mated pair each time they meet after being apart. Learn to recognize this display (*see* Courtship), for it will help you to identify mated pairs and understand other interactions.

It is also valuable to be able to distinguish the sexes. Since there are no differences in their plumage, you must rely primarily on their differences in behavior. The calls of the two sexes are very distinct. The male's is low, with two syllables: *ahonk*; the female's is higher, with usually only one syllable: *hink*. A truly amazing part of the Greeting Ceremony is that the male and female alternate their calls in such a well-timed way that the whole performance sounds as if it were given by just one bird. Sometimes when a pair fly overhead you will hear one call and then the other, and this will help you distinguish the two sounds.

Watch carefully for the various postures of the head and neck, for they are very expressive in Canada Geese and an important part of all the birds' visual displays. I once had a chance to test my ability to read Canada Goose displays when I was watching two pairs of Geese on a pond — each in its own territory. I had heard that one pair was tame but the other was not. While I was sitting and watching, one of the males started walking directly at me. Canada Geese are very aggressive defenders of their territories, and if you approach too closely they will charge and can inflict a bad bruise with their beak or wings. I didn't know if this was the tame Goose, but since it had made no aggressive displays before approaching, I decided to stay put. It was a true test of my belief in animal displays, and I was actually quite nervous as it got close. It stopped a few feet from me and began grazing the weeds and grasses — it was the tame Goose. As I moved about the lake, the other male spotted me and began to do all of the typical aggressive displays of Head-pumping and Head-forward, so I kept my distance and the displaying stopped.

BEHAVIOR CALENDAR

	TERRITORY	COURTSHIP	NEST-BUILDING	BREEDING	PLUMAGE (MOLTS)	SEASONAL MOVEMENT	SOCIAL BEHAVIOR
JANUARY							
FEBRUARY							
MARCH							
APRIL							
MAY							
JUNE							
JULY							
AUGUST							
SEPTEMBER							
OCTOBER							
NOVEMBER							
DECEMBER							

DISPLAY GUIDE

Visual Displays

Bent-Neck

Male or Female *Sp Su F W*

With the neck coiled back, the head is lowered and pointed toward the opponent.

CALL None or Hiss-call

CONTEXT Given in conflict situations with other Geese; usually considered a mild threat. *See* Territory

Head-Forward

Male or Female *Sp Su F W*

The neck is extended and the head is held low and pointed toward the opponent.

CALL Ahonk-call, Hink-call

CONTEXT Usually follows Bent-neck and is an expression of increased threat and incipient attack. *See* Territory, Breeding

Head-Pumping

Male or Female *Sp Su F W*

The head is rapidly lowered and raised in a vertical pumping motion.

CONTEXT Given in conflict situations and often precedes direct attack. *See* Territory, Breeding

Upright-Neck

Male or Female *Sp Su F W*

The neck is vertical and straight and the head is horizontal. The front of the body is tilted slightly upward.

CONTEXT Done by Geese that are wary of possible danger

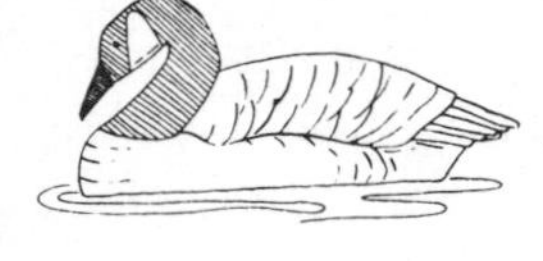

Head-Down

Male or Female *Sp Su F W*

The neck and head are drawn into the breast and the bill is pointed straight down.

CONTEXT Posture assumed by a Goose when near more dominant Geese; believed to be a display of submission that may appease aggression in the more dominant birds

Head-Flip

Male or Female *Sp Su F W*

With the neck held straight up, the bird flips its head from side to side.

CALL Quiet grunts

CONTEXT Given when a bird is slightly disturbed or apprehensive; also given just before it is about to move, either to walk or to take flight

Rolling

Male or Female *Sp Su F W*

The neck is extended horizontally and the head is rotated as in the Head-flip, only more vigorously. The head and neck are waved back and forth in an arc.

CALL Ahonk-call, Hink-call

CONTEXT May evolve from Head-pumping

display; given toward other males as an intensely aggressive display, and toward the mate as part of the Greeting Ceremony. *See* Territory, Courtship

Neck-Dipping

Male or Female *Sp*

The bird rhythmically and repeatedly dips its whole neck deeply into the water. This is often done with another bird.

CONTEXT A premating display that is usually done between a male and a female but may be done between two birds of the same sex. *See* Courtship

Head-Up

Male or Female *Sp*

Two birds, with their necks extended, lift their heads up at a slight angle.

CONTEXT Occurs after mating and is done simultaneously by both birds. *See* Courtship

Auditory Displays

Ahonk-Call

Male *Sp Su F W*

A slightly lengthened two-syllable call, usually repeated with space in between; second syllable higher than the first; a typical Goose *honk*

ahonk. ahonk. ahonk.

CONTEXT Given by the male during aggressive encounters, during the Greeting Ceremony, and at times when he is calling his mate. *See* Territory, Courtship, Breeding

Hink-Call

Female *Sp Su F W*

hink. hink. hink.

Higher and shorter than the male's Ahonk-call; sounds much like the bark of a small dog

CONTEXT Often alternated with the male's Ahonk-call during the Greeting Ceremony, making a duet so perfectly timed that it is hard to believe two birds are calling. *See* Courtship

Snore-Call

Male *Sp Su F W*

A low, drawn-out, grating call much like a human snore

CONTEXT Given by the male during the Greeting Ceremony as his head and neck are held low to the ground; alternated with the Ahonk-call and the Rolling display. *See* Courtship

Hiss-Call

Male or Female *Sp Su F W*

A hissing sound unlike any other call of the Geese

CONTEXT Given by the male or female at times when they are defending their territory, their nest, or their young; usually given only at close distances

BEHAVIOR DESCRIPTIONS

Territory

Type: Mating, nesting, feeding
Size: Variable, ¼–1 acre
Main behavior: Ahonk-call, Rolling, direct attack, Greeting Ceremony
Duration of defense: From the moment of nest site selection to the time the eggs have hatched

The territorial behavior of Canada Geese is one of the most exciting features of their lives to observe, for the birds have well-defined territories that are conspicuously defended. As soon as a nest site has been chosen, the male begins to defend a sizable area surrounding it while the female defends only the nest site. When another Goose approaches the territory, the male gives the Rolling display accompanied by loud bursts of the Ahonk-call. In most cases this is enough to make the intruder leave, but if he doesn't, then the male attacks him with neck extended and head held low. Fights can be prolonged and severe if the intruder is dominant enough to keep challenging the territorial male. The best way to see these territorial disputes is to listen for the loud Ahonk-call of the male and then to go quickly to the area it is coming from.

Other territorial displays of Geese include: Head-pumping, Bent-neck, Upright-neck, and Head-flip. These occur during minor disturbances that do not call for full territorial defense.

Following any territorial challenges, the male and female do a series of displays together called the Greeting Ceremony (also called the Triumph Ceremony by some writers). This involves loud calls and conspicuous movements and is fully described under Courtship.

An amazing aspect of territoriality in Geese is how quickly the territory dissolves. Once the young have hatched and can move about, the male stops all defense of the area, and other families of Geese are free to enter it. If you have been following the actions of a territorial male through the incubation phase, then this sudden change in his behavior is a striking occurrence.

Courtship

Main behavior: The male defending the immediate area around the female; the female closely following the male of her choice; the Greeting Ceremony
Duration: First pairing usually in winter; pairs remain together for as long as both Geese live.

The first signs of pair formation in Canada Geese are seen in the large winter flocks. Females show their choice of a mate by following him closely on land and in water. The male shows his choice of a mate by defending the immediate area around her from the intrusion of other males. Once the two are committed to each other, they frequently perform the Greeting Ceremony.

The Greeting Ceremony occurs between mates or family members when they have just come together after being apart. Between mates, the ceremony is most often given just after a male has had an aggressive interaction with another male and returns to the female, or when the female leaves the nest to feed and is joined by her mate. In the ceremony the male alternates the Snore-call with the Ahonk-call and its accompanying Rolling display. At the same time, the female holds her neck in a diagonal position and gives her Hink-call.

A feature of the display that is hard to believe even after you have heard it is that the male Ahonk-call and the female Hink-call are given as a duet. Each alternates with the other in such a perfectly timed way that together they sound like one call. In fact, what most of us think of as the call of the Canada Goose is actually a duet of the male and female calls. When a pair that is calling flies overhead, it is often possible to distinguish the two calls as first one passes over and then the other.

Displays given during mating are conspicuous. They are not commonly seen among wild Canada Geese, but if you visit any urban duck pond you are likely to see very similar displays given by domestic Geese. The main precopulatory display is Neck-dipping. In this the birds rhythmically dip their heads and necks deeply into the water and toss water over their backs as they bring them out. During the display the male aligns himself alongside the female and finally steps onto her back while holding her neck feathers in his bill. After copulation, both birds do Head-up. Throughout the Neck-dipping phase, both birds tread water with their feet, stirring it up on the sides.

Canada Geese remain paired for as long as both live. If one dies then the other usually finds another mate in a year or less.

Nest-Building

Placement: On the ground at the edge of open water or on small hummocks
Size: Outside diameter of 16–20 inches
Materials: Cattail leaves, grasses, and lining of feathers

The search for a nest site begins before territory is determined. The female takes the lead and the male follows, as the two birds swim about in suitable breeding areas. When they come to a possible spot, the female climbs up and explores it. If this spot is within another Goose's territory or is even already occupied by another nesting female, the male takes the lead and tries to enter the territory and force the female off. If he is successful, the intruding pair will nest there; if not, they will explore further.

Once a nest site is established the female begins building. She hollows out a shallow impression in the earth and gathers all available material that can be reached while standing in the nest. If more material is needed, it is broken off from near the nest and brought back. The female builds the nest around herself as she sits or stands in the shallow depression. It takes her about four hours to finish it.

Locating the Nest

WHERE TO LOOK At the edges of protected ponds, small lakes, or swamps; often on rocks or grass hummocks out in the water

WHEN TO LOOK In April or as soon as the first single pairs arrive in possible breeding areas

BEHAVIORAL CLUES TO NEST LOCATION:

1. Listen for loud honking, which may signify a Greeting Ceremony or territorial squabble. The nest will be in the area.
2. Watch for pairs quietly exploring with the female in the lead. This is a nest site search.
3. Look for a lone male feeding or resting who is aggressive to other Geese or to you. Its mate is on the nest nearby.

Breeding

Eggs: 5 eggs average
Incubation: 28 days, by female only
Nestling phase: Only 1 day, or none
Fledgling phase: 2–3 weeks; not clearly defined in Geese
Broods: 1

Egg-Laying and Incubation

The first egg may be laid within an hour after the completion of the nest. The remaining eggs are laid one per day until the clutch is complete. After a new egg is laid, the female leaves the nest, but before doing so, she covers it over with leaves or feathers. Full-time incubation starts around the time the last egg is laid and is done by only the female. She leaves the nest only two or three times per day and for periods of from just a few minutes to an hour or more. Often when the female leaves the nest, the male joins her, and they perform the Greeting Ceremony. Incubation lasts for twenty-eight days.

Nestling Phase

Young Geese are ready to leave the nest almost as soon as they hatch. Generally they are brooded for the whole first day. For the next few days, they are brooded only at night and for a few hours each day; the rest of the time they swim about with the parents. All brooding stops after the first few days.

Fledgling Phase

Throughout the fledgling phase the young can walk, swim and feed on their own, and are dependent on their parents only for protection. The family moves about as a unit, staying close together at all times. It roams about freely in areas where there is adequate food and protection, and this often brings it into contact with other families of Geese. Territorial boundaries have dissolved, but each family still keeps its distance from the others. When two families get too close, the adults start bouts of displays, including Upright-neck, Head-pumping and Head-forward, all in complete silence — a striking contrast to territorial defense, which is very loud. These displays make the two family groups move apart.

Plumage

Adult Canada Geese go through one complete molt per year. This occurs in midsummer and is completed by about mid-August. The young are still with the adults during the molt, and at this stage none of the family can fly, the young because they haven't grown their full wing feathers and the adults because they are molting their wing feathers. Thus the birds move to more secluded areas and resort to elusive behavior when threatened. In late summer all of the family can fly, and they move to open areas where there is abundant food and join with other Geese to form large flocks.

When the male and female are together you can usually tell the male, for he is slightly larger. Other than this you must rely on behavior to distinguish the sexes, and in the case of Geese this is quite easy. The male is the only one to give the deep Ahonk-call and the Snore-call; he is also the first to defend the territory. The female has the much higher Hink-call, does all of the incubating, and normally does not participate in defending the territory.

Seasonal Movement

Canada Geese are generally migratory, moving in conspicuous lines or V-shaped formations high in the air as they travel south in fall and north in spring. Over the last twenty years this pattern has changed slightly due to wildlife management practices of providing food for the Geese throughout the winter. Now many flocks of Geese are remaining in northern areas in winter, and the migration patterns of the species are not as clear-cut as they used to be.

FL

Mallard / *Anas platyrhynchos*

MALLARDS ARE ONE OF THE HIGHLIGHTS OF WINTER BEHAVIOR-watching, for their most active courtship starts in fall and continues right on until spring. Mallard courtship consists primarily of displays done among groups of males and displays done between a male and a female. The group displays are the most exciting, for in them three or more males perform intricate movement patterns almost simultaneously. Although commonly given, the displays are still hard to see, for they are subtle and last only a few seconds. The Head-shake and Tail-shake displays usually precede the more complicated displays, and can be used as indicators of where you should direct your attention. If you look over the display guide and read the behavior description of courtship before going out to observe, you will have a better chance of seeing the displays.

When I first started studying Mallards I was surprised to find that the males and females make entirely different sounds. The *quack*ing sound, which I had assumed all Ducks made, can be made only by the female. The male has two other calls of his own — a nasal *rhaeb* sound and a short Whistle-call, the latter accompanying all of the group courtship displays. This last call is particularly interesting, for it is so unlike the sounds we assume Ducks make.

An added advantage to knowing Mallard displays is that closely related species of Ducks such as Black Ducks, Gadwalls, Pintails, Widgeons and Teals have many similar displays. Therefore, once you learn some of the patterns of Mallard behavior, you will have a good start on being able to understand the behavior of these other Ducks as well. The Black Duck is particularly close in this respect, having nearly the same display repertoire as the Mallard.

Mallard territory is interesting, for it does not include the nest, but is in fact usually quite far from it. It is only defended for a short time before incubation starts, and probably functions to isolate the male and female during mating. Discovering other aspects of Mallards' lives besides courtship can be very challenging, for the birds are quite secretive during breeding and may nest in places that are inaccessible to the average observer.

BEHAVIOR CALENDAR

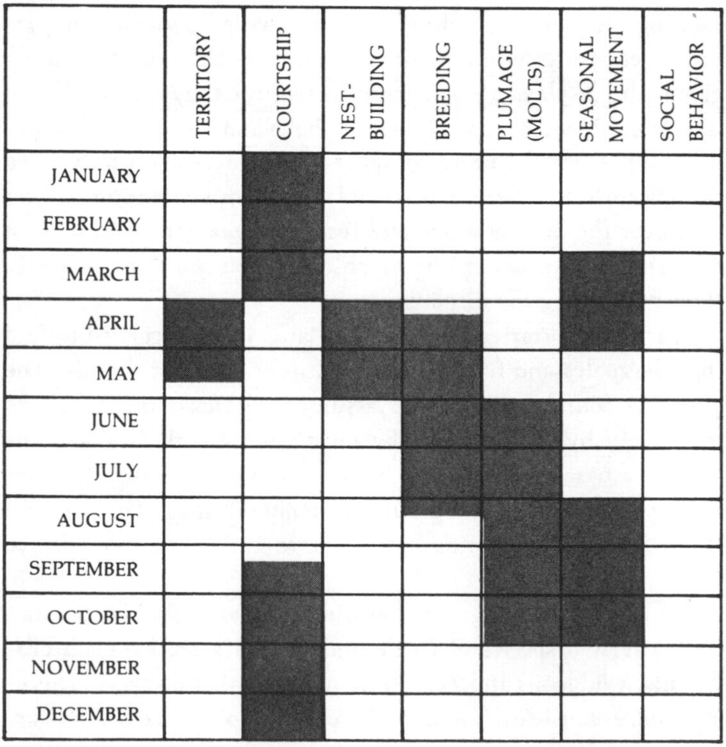

	TERRITORY	COURTSHIP	NEST-BUILDING	BREEDING	PLUMAGE (MOLTS)	SEASONAL MOVEMENT	SOCIAL BEHAVIOR
JANUARY		■					
FEBRUARY		■					
MARCH		■				■	
APRIL	■		■	■		■	
MAY	■		■	■			
JUNE				■	■		
JULY				■	■		
AUGUST					■	■	
SEPTEMBER		■			■	■	
OCTOBER		■			■	■	
NOVEMBER		■					
DECEMBER		■					

DISPLAY GUIDE

Visual Displays

Head-Shake and Tail-Shake

Male *F W Sp*

Both of these displays start with the head held close to the body, concealing the white neck ring. In Tail-shake the tail is given a brief shaking. In normal Head-shake the bill is just shaken from side to side; in more intense versions the bird raises its breast out of the water and stretches its neck as it shakes its head. The wings are not flapped as they are in the similar bathing movement.

CONTEXT Given before other more elaborate courtship displays, and may serve to draw the courted female's attention to the coming displays. *See* Courtship

Grunt-Whistle

Male *F W Sp*

The bird raises the back of his neck while keeping his bill pointed down. A whistle is given when the neck is most arched, and just before this an arc of droplets is tossed into the air. The whole display may take less than a second to complete.

CALL Whistle-call

CONTEXT Given by small groups of males, often simultaneously, in front of a female. *See* Courtship

Down-Up

Male *F W Sp*

The bird tips forward to dip his bill and breast into the water, then lifts only his bill and gives a Whistle-call followed by a Rhaebrhaeb-call. The display lasts no longer than two seconds.

CALL Whistle-call, Rhaebrhaeb-call

CONTEXT Used in small groups with other males, and often given simultaneously with their displays. *See* Courtship

Head-Up-Tail-Up

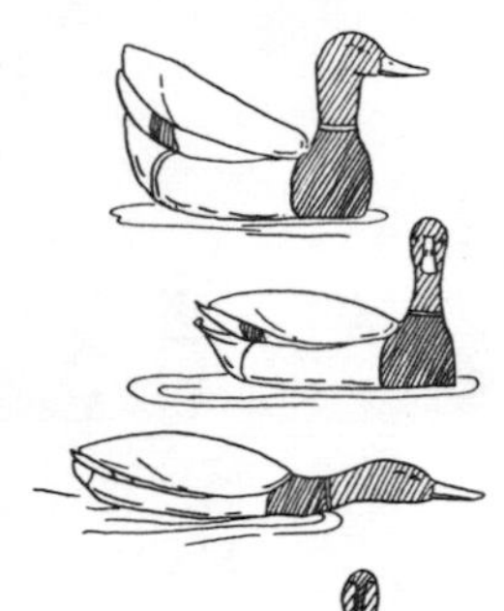

Male *F W Sp*

In this rapid series of movements, the male stretches his head, wingtips and tail upward, then lowers his wings and tail and turns his head toward a female. He next gives a short display of Nod-swimming, and turns his head away from the female.

CALL Whistle-call

CONTEXT Given in the presence of other males and at least one female. *See* Courtship

Inciting

Female *F W Sp*

The female follows behind a male while repeatedly flicking her bill back over one side of her body. The accompanying call is distinctive.

CALL Inciting-call

CONTEXT Used by females as they follow their mates and are challenged by other males; very common. *See* Courtship

Nod-Swimming

Male or Female *F W Sp*

With its neck outstretched and its head just grazing the water, the bird swims rapidly for short distances.

CONTEXT Males Nod-swim in the Head-up-tail-up display and immediately after mating. Females Nod-swim near groups of males, and it often causes the males to do other displays. *See* Courtship

Mock-Preening

Male and/or Female *F W Sp*

One wing is lifted to a vertical position and its feathers are fanned. This exposes the row of blue feathers on the wing (the speculum). The bird then places its bill behind the fanned wing as if it were preening.

CONTEXT Often given simultaneously by a male and a female, and may signify a strengthened commitment between the two birds. *See* Courtship

Pumping

Male and Female *W Sp*

A male and a female face each other and begin rhythmically bobbing their heads up and down.

CONTEXT Usually followed by mating, but may just slow down and stop; sometimes done between two males. *See* Courtship

Auditory Displays

Whistle-Call

Male — *F W Sp*

A short, high whistle; not a sound the beginning behavior-watcher would expect a duck to make

CONTEXT Given only during the displays: Grunt-whistle, Down-up, and Head-up-tail-up. Listening for this call is the best way to catch the displays, as they are so fast. *See* Courtship

Slow-Rhaeb

Male — *Sp S F W*

rhaaaeb. rhaaaeb.

A drawn-out, nasal sound, usually given a number of times with a pause after each call

CONTEXT Given by males when there is danger or a disturbance nearby

Rhaebrhaeb-Call

Male — *Sp S F W*

rhaebrhaeb, rhaebrhaeb,

A short, nasal, two-syllable call, often quickly repeated a number of times

CONTEXT Given during aggressive encounters between males, and may continue for a half minute or more

Inciting-Call

Female — *F W Sp*

quegegegegegege

A rapid series of short, percussive notes

CONTEXT Given only during the Inciting display, and is often the best way to locate the display among a group of ducks. *See* Courtship

Decrescendo-Call

Female *Sp S F W*

A series of about seven resonant *quacks* given with a slight space after each one and decreasing in loudness and pitch, the second *quack* usually being the loudest

qua QUA qua qua qua qua qua

CONTEXT Given by the female when she is uneasy or temporarily separated from her mate

Persistent-Quack

Female *Sp Su*

A series of well-spaced *quacks* given at the same intensity and over a long period of time

quack. quack. quack.

CONTEXT Done by the female when she is looking for a nest site; may be given while on the water or in the air. *See* Nest-building

BEHAVIOR DESCRIPTIONS

Territory

Type: Feeding, mating
Size: 1/8–1/4 acre
Main behavior: A lone pair seen in the same small area for over a week; some defensive behavior on the part of the male
Duration of defense: 10–14 days; ends as incubation starts

From the start of nest-building until the first few days of incubation, the male defends a small area of open water and reeds. This is separate from the nest site but usually no more than one hundred yards from it. When the female is not occupied at the nest, either building, egg-laying, or incubating, she joins the male in the territory.

Mallard territory is inconspicuous, for it does not contain the nest, is not advertised by the male, and only occasionally needs to be defended. Three clues will help you to recognize a Mallard territory. First of all, you will see either just the male or the pair in the same area day after day for about two weeks. Also, if for some reason the birds are flushed into flight, they will return to the same spot when the danger is past. And last, they do not allow other Mallards on their territory, except on some occasions when first-year males are allowed in. Aggressive defense of the territory occurs only in areas where there is a large breeding population and competition for space.

The territory generally includes open water for feeding, dense cover for protection, and an open area of land called a "loafing area" for preening and resting. Around the edge of a lake or pond the scarcity of loafing areas is often the main factor limiting the number of territories.

Courtship

Main behavior: Many group displays, pair-formation displays, matings
Duration: From September to April

Courtship in Mallards occurs from fall to early spring, a time when the Ducks congregate in large flocks. The most common activities in fall and winter are group displays and pair formation. Later in spring, matings and forced matings are more likely to be seen.

Group Displays

The first sign of competitive courting is two or more males milling about each other on the water with their heads drawn in close to their bodies. From this position they will do the displays of Head-shake or Tail-shake. Following these preliminary displays the males may do one of three different displays: Grunt-whistle, Head-up-tail-up, or Down-up. This activity is always in the presence of one or more females, and each male directs his displays to a particular female that he is courting. All three of the final displays are accompanied by the short Whistle-call, which is very conspicuous and can draw your attention to displaying birds. The displays themselves are from one to a few seconds long, and there is no time to have a display pointed out — you must be looking at the birds as they display. The preliminary Head-shake and Tail-shake are helpful in that they warn you which birds are likely to be seen displaying. They may also attract the female's attention to the displaying male. The female has one display that is obvious among these group displays: Nod-swimming, where she rapidly swims among a group of males, with her head low to the water. This always causes the males to go through a burst of displays immediately afterward.

Group display for Mallards reaches its peak in November and December, then increases again slightly in February and March.

Pair Formation

Pair formation is different from the group displays in that it involves just a male and female and their mutual signs of accepting each other. The most obvious of these is the female's Inciting display, which is used as she follows her chosen mate.

The call that accompanies this display is distinctive, and is the best way to locate the visual display. It is unique to the female, and has been aptly described as a tremulous *quegegegegegege*. It is activated by the approach of a strange male, whereupon the female swims closely behind her mate and flicks her bill back to one side or the other. It is extremely common and can be seen in water, on land, or in the air.

The male often responds to the female's Inciting by a subtle display of his own called "leading," where he turns the back of his head toward the female and swims slowly away. It may be that this stimulates her to follow him. Another pair-formation display is Mock-preening, which utilizes the bright blue speculum on the Mallard's wings. The bird places its bill behind one wing and then spreads the wing vertically like an opening fan, holding the position for a number of seconds and orienting its spread wing toward the other bird. Although not as common as the others, this display is perhaps the most visually striking.

When a pair become separated, the female will often give the Decrescendo-call: *qua QUA qua qua qua*. When rejoining after a separation or a skirmish with other birds, they may participate in Rhaebrhaeb palaver. Here the two birds remain near each other and the male repeatedly gives the Rhaebrhaeb-call while lifting his bill slightly above the horizontal.

Mating

Mating between paired Mallards is almost always preceded by Pumping, where the two birds face each other and repeatedly jerk their heads downward. This display has to reach a certain intensity before mating takes place, and when this is not the case, the display just stops. If it reaches the right intensity, the female begins to lower her head closer and closer to the water. At this point the male steps onto her back, grabs the feathers on the back of her head with his bill, and makes cloacal contact. He then gets off, pulls his head back quickly (called "bridling") and then Nod-swims around the female while she bathes.

In spring, Pumping may be seen between pairs of males, and it occasionally leads to mating attempts.

Forced Mating

In late spring and early summer some males try to forcibly mate with other females. Unlike matings between paired birds, these involve no preliminary displays. The male simply chases the female, steps on her back, takes her head feathers in his beak, and copulates. These are particularly common during the time of incubation and during the fledgling phase. The female usually gives an aggressive call when confronted by these males, but this often does little to stop them. Aerial chases may result, the female dodging in flight with one or more males behind her.

Nest-Building

Placement: On the ground near water, in woods or brush
Size: 8 inches inside diameter
Materials: Grasses, leaves and reeds

Nest-building begins with the pair's flying together over suitable nesting locations. The female is always in the lead and it is she who explores the specific sites while the male waits nearby. The nest site and the territory are separate but usually within two hundred yards of each other. Nest searching may go on for a week or more, most of it occurring in the early evening.

This period in the Mallard's life is marked by the Persistent-quack of the female. It is a series of single monotonous quacks at the rate of about two per second and is accompanied by extreme alertness. It starts a week or more before egg-laying, then becomes less common after the first egg is laid.

The nest is built entirely by the female. At first it is just a few scrapes in the ground, but as the eggs are laid, the female spends more time in building it. At best, it is just a collection of leaves, grasses and reeds from the immediate area, forming a thick rim of plant material that fits snugly against the contours of the female's body. During incubation, down from the female's breast is added to the nest's lining, and this is drawn up over the eggs whenever she leaves.

The birds will nest in many different surroundings: in open fields, in tall grasses, or even on muskrat homes.

Locating the Nest

WHERE TO LOOK In woods, fields, cattails, near water; may be placed 100 feet or more from the water's edge, but is usually closer

WHEN TO LOOK Early spring

BEHAVIORAL CLUES TO NEST LOCATION:

1. Listen for and locate where Persistent-quacking is taking place.

2. Watch for lone females in nesting habitats.
3. Locate territories and watch for activity near them.

Breeding

Eggs: 8–10
Incubation: Average 23 days, all by female
Nestling phase: 1 day or less
Fledgling phase: 50–60 days
Broods: 1

Egg-Laying and Incubation

The female lays her eggs in the early morning, usually one per day until the clutch is complete. In between her periods of laying she joins the male on the territory. Once the clutch is complete she starts to incubate full time with only one or two brief visits to the territory each day. During these visits she drives the male away from her, and he ceases to defend the boundaries of this area. When she leaves the nest, she covers the eggs with grasses or dry leaves from around the area. The gesture of repulsion given to her mate is also her response to all males at this time, and strange males will only rarely force their attentions upon her. The average incubation period is twenty-three days.

Nestling Phase

The ducklings stay in the nest for most of the first day and are brooded by the female. As with the Canada Goose, there is no real nestling phase.

Fledgling Phase

The second day, the young are led by the female to the nearest water. Here they bathe, preen, and feed with her for the next fifty to sixty days. At the beginning of this period the ducklings are brooded by the female at loafing spots, but when they get bigger they sleep on their own beside her.

The female and her brood are very intolerant of other females and their broods, sometimes pecking them hard and even drowning them if they come too close.

At the end of this period, when the ducklings can fly, the female leaves them and goes to protected areas where she undergoes her complete molt.

Plumage

Mallards go through two molts per year. A complete molt of all feathers occurs in late summer and fall — it is during this molt that the male assumes his well-known appearance of a green head and light-colored body. In early to midsummer both the male and female go through a second molt; this is incomplete, involving only the body feathers and not the wings or tail. During this molt the male assumes plumage exactly like that of the female. This plumage is kept for only a month and is called the "eclipse" plumage. This is the one time of the year when it is difficult to distinguish male from female Mallards. The male starts this molt in early summer usually after leaving the incubating female, while the female begins hers later, most often after her young are independent. The female's appearance does not change with either molt.

Most birds assume their brightest plumage in spring, but the male Mallard is most colorful in fall and winter. This makes sense to the behavior-watcher, for this is when Mallards are going through their most active phase of courtship, while for most other birds this occurs in spring.

For most of the year the male and female can be easily distinguished by their appearance: the male has a green head and white neck ring, and the female is streaked brown all over. But the behavior of the sexes also differs significantly, the female being the only one to give the typical *quack*ing sounds while the male gives nasal *rhaeb* calls and the Whistle-call. Also, Inciting is done by only the female, while Head-up-tail-up, Grunt-whistle and Down-up are done by only the male.

Seasonal Movement

Mallards are generally migratory, most flying south to the Gulf Coast in fall and back to the northern United States and Canada in spring. But many Mallards remain north throughout winter in areas where there is sufficient food and some open water.

In late August and early September, flocks of Ducks will be seen moving about. This is a seasonal movement of the Ducks as they move from their breeding grounds to large lakes and marshes for feeding. Actual migration occurs about a month later.

American Kestrel / *Falco sparverius*

AMERICAN KESTRELS ARE PRIMARILY SOLITARY BIRDS. EVEN DURING the breeding season, each adult spends most of its time alone. This solitude results from the marked division of labor between the sexes of a breeding pair. The male remains on the hunting grounds, catching prey for both himself and the female. The female always remains in the immediate area of the nest, resting, preening, incubating the eggs, or feeding on prey brought to her by the male. This extreme division of labor can last for eight to twelve weeks, for it includes a number of weeks before incubation, the incubation phase, and the first few days of the nestling phase.

The majority of Kestrel displays occur during the male's transfer of food to the female and since mate-feeding continues for such a long period, you have a good chance of being able to see them. In general, the food is brought to the nest area by the male, who approaches with a flight display and several calls. The female flies out to him, also giving several calls, and follows him for a short distance. They land at a special perch, and the food is transferred as the two bob their heads up and down. The female may then carry the food to her own special perch, where she may eat some of it and store the rest. The male goes off to continue hunting. It is interesting that this pattern begins long before the female is actually involved with egg-laying or incubation; in other words, it starts when she is still perfectly capable of hunting for herself.

There are only three basic calls of the American Kestrel, and they are fairly easy to distinguish. The one most people hear is the Klee-call, a generalized auditory display given in situations where something excites or disturbs the bird. The other two are slightly more specific. The Whine-call is given usually during food transfers, such as those between mates or between adults and young. The Chitter-call is most often given during other types of close interactions between members of the family: when the birds copulate, when one approaches the nest, and during certain moments of the mate-feeding ceremony.

BEHAVIOR CALENDAR

	TERRITORY	COURTSHIP	NEST-BUILDING	BREEDING	PLUMAGE (MOLTS)	SEASONAL MOVEMENT	SOCIAL BEHAVIOR
JANUARY							
FEBRUARY							
MARCH							
APRIL							
MAY							
JUNE							
JULY							
AUGUST							
SEPTEMBER							
OCTOBER							
NOVEMBER							
DECEMBER							

DISPLAY GUIDE

Visual Displays

Flutter-Glide

Male or Female *Sp Su*

The bird flies with outstretched wings, arched below the horizontal plane of the body, and rapidly flutters them.

CALL Whine-call, Chitter-call

CONTEXT Occurs during mate-feeding as male approaches with food and as female follows him; also occurs as male shows nest site to female, and as he approaches the female for mating. *See* Courtship, Nest-building, Breeding

Dive

Male *Sp*

The male does a series of deep dives and upward swoops over the nest area. The wings are beat deeply throughout the dives and at the peak of each ascent a few Klee-calls are given.

CALL Klee-call

CONTEXT Done near the nest site, occurring most during the early phases of courtship. *See* Courtship

Auditory Displays

Klee-Call

Male or Female *Sp Su F W*

A series of sharp staccato notes, similar to the written sound *kleekleekleeklee*

CONTEXT Given during aggressive encounters

with other birds or predators near the nest or in the territory; also given during the Dive display. *See* Territory, Courtship, Breeding

Whine-Call

Male or Female *Sp Su*

A loud, continuous sound rising in inflection; sometimes combined with Chitter-call; may be repeated for as long as 1–2 minutes
CONTEXT Occurs most in relation to the exchange of food: given by both male and female during mate-feeding, and by the fledglings when begging for food. *See* Courtship, Nest-building, Breeding

Chitter-Call

Male or Female *Sp Su*

A harsh chattering call, given in rapid bursts; sometimes combined with the Whine-call
CONTEXT Given in close interactions with a mate or young, i.e., during mating, mate-feeding, nest-relief. *See* Courtship, Nest-building, Breeding

BEHAVIOR DESCRIPTIONS

Territory

All American Kestrels have a breeding territory that they occupy during summer, and in at least some areas of the country they also have a territory in winter for the purpose of protecting a hunting area.

Breeding Territory

Type: Mating, nesting, feeding
Size: Average 250 acres, variable
Main behavior: Chase, Klee-call
Duration of defense: March through June

Territorial defense is not a conspicuous aspect of Kestrel behavior. Males generally arrive on the breeding site ahead of the females and localize their activities to several hunting areas and a possible nest site. After they pair with a female they restrict their activities even more to an area averaging two hundred fifty acres (in mountain regions). This size will vary tremendously with the habitat and the availability of food. Neighboring males rarely have conflicts and seem to avoid each other's areas. However, foreign males flying through a resident male's territory will be chased out by the resident. The Klee-call often accompanies this chase.

Nonbreeding Territory

Type: Feeding
Size: Over 100 acres, variable
Main behavior: Chase, Klee-call
Duration of defense: August through January (approximate)

In some areas Kestrels also defend winter territories, which are their hunting grounds. A pair residing all year either remains together on its breeding territory, or creates separate territories near to each other. The competition occurs when migrants from the north or higher altitudes enter these areas to spend the winter. A Kestrel chooses its territory in early fall and defends it until breeding activities begin in late winter or early spring.

One unusual aspect of Kestrel behavior is that females and males seem to concentrate in different habitats throughout winter. Some studies have suggested that the females occupy the more favorable hunting areas, those with more open land, and that the males winter in areas with dense growth or at the edges of urban areas.

Courtship

Main behavior: Copulation, mate-feeding, aerial dives, Klee-call, Whine-call, Chitter-call
Duration: 3–4 months; variable, possibly depending on climate

The two most common aspects of courtship seen in the American Kestrel are mate-feeding and mating. The mating behavior pattern generally begins when the female gives the Whine-call. At this signal the male flies to her (with Flutter-glide) and lands on her back. The female moves her tail to one side as the male sharply lowers his tail several times to make contact with her. The male may have his wings slightly outstretched for balance, and he may give the Chitter-call. In a few seconds the male flies to a perch nearby, and both birds begin a short period of preening. Your attention will be drawn to this behavior by the loud Whine-call and Chitter-call, given in various combinations by either sex. Occasionally, mating occurs quietly, with the male flying directly onto the female's back with no call from her, then leaving quietly himself as well.

Mating can occur before pair formation, when the females move freely in and out of male territories. This happens early in the season; later they restrict their movements to their mate's territory. Mating can occur as frequently as fifteen times a day and over a period of six weeks. Its frequency decreases once egg-laying has started.

Mate-feeding begins only after two birds have become paired for the breeding season. When the female first starts to remain on the male's territory, the two birds may hunt together. But soon she stops all of her hunting, remains near the nest, and has all of her food brought to her by the male. This behavior continues until one or two weeks after the young have hatched — a period of up to eleven weeks.

The transfer of food is accompanied by several displays and is such a conspicuous behavior that if there are Kestrels anywhere in your area you should be able to observe it. As the male, with prey in his talons, nears the vicinity of the nest, he starts Flutter-glide and gives the Whine-call, sometimes in combination with the Chitter-call. For the last part of the flight the female may fly out and follow him with the same visual and auditory displays. The male lands at a fixed food-transfer spot and the female lands next to him. Both birds bow or bob their heads several times. Then, as the female takes the food, both birds chitter and flutter their wings. Following this, the female takes the food to one of her perches, and the male leaves to continue hunting. The female may then eat it or cache it in the top of a broken limb or in the crotch of two branches. This is the complete behavior pattern, but often only parts of it are seen in any given transfer.

Another behavior possibly related to courtship is the aerial dive of the male. In this the male repeatedly flies up and then dives steeply. It is usually accompanied by the Klee-call and is generally not a common behavior pattern.

Nest-Building

Placement: In tree holes previously excavated by other birds, 15–30 feet high
Size: Nest hole opening is approximately 2–4 inches in diameter
Materials: Just the tree cavity, no lining material

You may have a chance to see Kestrels looking for a nest site. The female may explore nest holes on her own, but only those that she is already familiar with. Generally, however, the male takes the lead, and with the Flutter-glide, food-carrying, Whine-call or Chitter-call, seems to induce the female to follow him to the nest hole he is interested in. In many cases he will fly to the hole and perch at the entrance or enter briefly, and as the female arrives at the hole he flies to a perch nearby. The male may even interrupt his Flutter-glide when he approaches a female for mating, go briefly to the nest hole, Chitter-call, and then return to the female and complete the mating sequence. There are many variations on this type of behavior.

The Kestrels may have to compete for the nest hole, especially with Flickers or squirrels.

Locating the Nest

WHERE TO LOOK In older trees near open areas or just at the edge of the forest

WHEN TO LOOK Anywhere from February on, even though the eggs are not laid until late May (the pair establishes the nest site earlier)

BEHAVIORAL CLUES TO NEST LOCATION:

1. The conspicuous activities of food transfer and copulation both take place very near the nest.
2. Most vocal activity occurs near the nest.

Breeding

Eggs: 4
Incubation: 30 days, mostly by female
Nestling phase: 30 days
Fledgling phase: Approximately 14 days
Broods: 1

Egg-Laying and Incubation

Eggs are laid one to three days apart, and incubation starts slightly before the laying of the last egg. The female does the majority of incubating, being relieved normally only two times in the day — in the morning and late afternoon. The male's share of incubation generally is about four hours of each day. When the female comes off the nest she feeds and preens.

Nest-relief is generally accomplished through auditory displays. As the male approaches the nest with or without food, he Flutter-glides and gives Whine-calls and/or Chitter-calls. The female leaves the nest at this signal and either accepts the food the male has brought or feeds from caches. The male then flies to the nest with the Whine-call. When the female is ready to return to the nest she first gives the Whine-call, and at this signal the male leaves to continue hunting. She then enters the nest and resumes incubation. Both male and female develop brood patches, so both do real incubation.

During this period there is very little visible activity around the nest, and the vocalizations during nest-relief may be slightly subdued. Thus, locating the nest during this phase is more difficult than doing so either before or after it, when there is more coming and going to and from the nest.

Nestling Phase

The young hatch over a period of three to four days. The female broods them steadily for the first few days, but by the ninth day she broods them only at night. During this time of brooding, the male continues to bring all of the food, now for both the female and the young. However, once the female stops daytime brooding, she takes on the greater share of food gathering. As the adults fly back to the nest with prey, they often take advantage of rising winds to lift them to the height of the nest, and thus conserve their strength. This nestling phase lasts approximately thirty days.

You might think the nest would get quite dirty with such a long nestling phase. Fecal sacs are not carried from the nest, but the young do shoot their feces onto the upper walls of the cavity where they dry quickly. Also, dermestid beetles are always plentiful in the nest, and they devour the remains of uneaten prey.

Fledgling Phase

The young may leave the nest over a period of a few days. During the day they remain in trees near the nest, making short flights and being generally active; at night they return to the nest hole with the adult female. As the adults approach with food, the young give the Whine-call and flutter their wings. They may even go out to meet the parents in flight. The adults are silent during these food exchanges. This distinguishes the exchanges from food transfer between mates, which occurs earlier in the season.

The period of dependence upon the adults for food lasts about fourteen days.

Plumage

American Kestrels go through one complete molt per year. This molt occurs generally just after the time of egg-laying. The molt of the female has been found to start a few weeks earlier than that of the male. The molts continue over a period of three to four months and end about the beginning of September.

The male and female can be distinguished primarily through their plumage. The male's wings are blue gray and his tail feathers are unbarred, while the female's wings are rufous brown and her tail feathers are horizontally barred with fine black lines.

Seasonal Movement

The migratory status of American Kestrels seems to vary with different regions of the continent. The birds are definitely migratory in the most northern parts of their range. Farther south, in areas where hunting can still take place during winter, some birds move away and others remain — in these cases it is often the males that remain. In southern areas the birds may not move at all from their breeding grounds, or drifting may occur, with some birds moving to more favorable habitats for hunting.

Social Behavior

In certain parts of the country where Kestrels are particularly abundant, large groups of juveniles may be seen hunting all in the same area. This occurs in late summer when the young have become independent of their parents. There may be ten to twenty juveniles gathered together if the hunting area is sufficiently large and abundant with prey animals. This behavior has been reported by a number of observers, mostly in the more open western and midwestern areas of the country. These juvenile groups are broken up either by migration or by the entry of adults seeking winter territories, both of which occur in the fall.

Herring Gull / *Larus argentatus*

SINCE HERRING GULLS BREED PRIMARILY ON SECLUDED ISLANDS off the East and West coasts and along the shores of our larger inland lakes, not many people have a chance to observe their breeding behavior. At the same time, the birds are extremely common and are often seen outside their breeding ground. Therefore, they have been included in this guide, not only for the value of being able to compare their behavior with that of other common birds, but also in the hope that some people may be stimulated to visit Gull colonies for behavior-watching.

One of the simplest things you can do with Gulls, even away from the breeding ground, is to become more sensitive to their different calls. A common call during summer is the Long-call. This, like many of the Gull's other breeding calls, may best be recognized by seeing the body movements that accompany it. See the display guide for a complete description. Two calls that can be heard all year are the *kleew* call, a display used in many situations, and the Alarm-call, which sounds like a high-pitched *gagagaga*.

When spring comes, watch for changes in the way Gulls fly about. In winter the Gulls move about individually or in flocks, but in spring older birds that have bred before join with their mates before going to the breeding ground. Thus, at this time you will start to see Gulls moving about in pairs, the male and female usually staying within a few feet of each other throughout the day.

If you do get a chance to observe breeding behavior, try to

start watching at the beginning of the breeding season. This will give you a better chance of seeing the courtship displays of Gulls that are breeding for the first time. Also, be prepared with liberal amounts of time and patience, for Gulls are wary and fly up in alarm even when you are still quite far away. Gulls also have long periods of resting and preening when very little interaction or displaying takes place. But whether you get to a nesting area or not, look over the display guide and behavior descriptions, for Gull behavior is fascinating, and knowing it may give you some insights into the behavior of other birds.

BEHAVIOR CALENDAR

	TERRITORY	COURTSHIP	NEST-BUILDING	BREEDING	PLUMAGE (MOLTS)	SEASONAL MOVEMENT	SOCIAL BEHAVIOR
JANUARY							
FEBRUARY							
MARCH							
APRIL							
MAY							
JUNE							
JULY							
AUGUST							
SEPTEMBER							
OCTOBER							
NOVEMBER							
DECEMBER							

DISPLAY GUIDE

Visual Displays

Upright

Male or Female *Sp Su*

The bird's neck is stretched upward while the bill is usually horizontal.

CONTEXT Occurs during aggressive interactions. It is usually done by a territory owner at an intruder, in which case the bird usually approaches the intruder while doing the display. If the bill is pointing up, the bird may be in a submissive rather than an aggressive mood. *See* Territory.

Choking

Male and Female *Sp Su*

Two birds together lower their bodies, tip forward, and repeatedly make pecking motions toward the ground. They may scrape the earth with their feet at the same time.

CALL Choking-call

CONTEXT Done by a pair during courtship, or on their territory as they seek out a nest site or assert their presence to neighboring pairs. *See* Territory, Courtship, Nest-building

Grass-Pulling

Male *Sp Su*

Two birds near each other start to pull and tug at grasses or other plants near them. Occasionally some plant material is loosened and they both pull on it as in a tug-of-war.

CONTEXT Occurs between males at territorial borders when competition over those borders is keen. *See* Territory

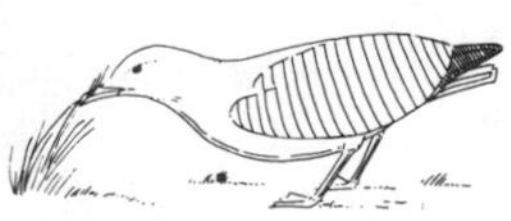

Head-Tossing

Male or Female *Sp Su*

Starting with its body in a horizontal plane and its head drawn in, the bird repeatedly flips its bill upward.

CALL Head-tossing-call

CONTEXT Done by males or females during courtship and may lead to mate-feeding or mating; also done in a slightly different manner by young birds when they are begging for food. *See* Courtship, Breeding

Head-Flagging

Male or Female *Sp Su*

The bird, with its body in the Upright display, suddenly moves its head to one side.

CONTEXT Usually occurs as a mutual display with both birds turning their heads away from each other; has been called "Facing-away" by some observers; usually a part of the mutual displays that take place in the early phases of courtship. *See* Courtship

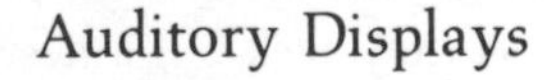

Auditory Displays

Alarm-Call

Male or Female *Sp Su F W*

gagagaga. gagagaga.

A brief series of short notes delivered rapidly; very much like the written description; usually three to six notes per series

CONTEXT Occurs in response to a possible predator on the breeding ground; may also occur in other situations of general alarm outside the breeding ground. *See* Breeding

Long-Call

Male or Female *Sp Su*

An extended call of roughly three parts and accompanied by specific bodily movements. First the head is arched forward, then it is moved under the breast, and finally it is tossed obliquely upward. The accompanying three parts to the call are similar to the written descriptions.

ow. ow. ow. kee, kee, kee, kyowkyowkyow

CONTEXT A very common call throughout the breeding season; can be recognized by the distinctive bodily movements that accompany it; may be used in courtship or aggressive encounters. *See* Courtship

Choking-Call

Male or Female *Sp Su*

A muffled, rhythmic series of sounds best recognized by seeing its accompanying display of Choking; sounds much like its written description

huoh huoh huoh

CONTEXT Occurs during the Choking display. *See* Territory, Courtship, Nest-building

Head-Tossing-Call

Male or Female *Sp Su*

A series of clear notes each accompanied by an upward tossing of the head; most easily recognized by the accompanying movements

CONTEXT Given by adults during courtship, or by young as they seek food from the parents. *See* Courtship, Breeding

Mew-Call

Male or Female *Sp Su*

A drawn-out, plaintive call given once or several times in succession; best recognized by the accompanying action of the head and neck arching downward and the bill opening wide

CONTEXT Usually occurs during courtship activities and may be done as a mutual display by a pair. *See* Courtship

Call-Note

Male or Female *Sp Su F W*

kleew kleew The typical call of the Gull heard throughout the year; sounds much like its written description

CONTEXT Does not seem to be associated with any particular circumstances

BEHAVIOR DESCRIPTIONS

Territory

Type: Mating, nesting
Size: 30–50 yards in diameter
Main behavior: Upright, Grass-pulling, Mew-call, Choking, chases and fights
Duration of defense: From arrival on the breeding ground until the end of the breeding season

The first birds to arrive on the breeding grounds are generally older birds that have previously mated. They arrive with their mates and most often occupy the same territory they have used in past years. These territories are situated in the best areas and form the center of the breeding colony. Younger birds usually arrive later and, after pairing, must claim territories. The easiest places for these new birds to create territories is around the periphery of the established colony, for at these spots the density of Gulls is low, and there is little competition. When they try to create territories within the established colony they have much more trouble, and aggressive encounters may occur.

Before egg-laying starts, it is difficult to follow the territorial activities of Herring Gulls, for the birds spend only part of each day on their territories — the rest of their time is spent in feeding away from the breeding ground. It is common for Gulls to intrude temporarily on a foreign territory in the absence of the owners.

There are three ways Herring Gulls deal with territorial conflicts, depending on the status of the owner and the intruder. The owner handles minor encroachments by doing the Upright display and walking toward the intruder. This may evolve into a short aerial chase ending at the territorial border, whereupon both birds face each other with upheld wings. In more intense conflicts, two territorial birds may meet at a common border and do Grass-pulling. This may lead to a fight if neither bird is about to give way. Newly paired birds defend territory in a third way, by walking around it together and stopping to give the Mew-call or do Choking. This may stimulate a neighboring pair to do Choking at the same spot.

Courtship

Main behavior: Long-call, Mew-call, Choking, Head-flagging, mate-feeding, Head-tossing
Duration: From the time the birds break from winter flocks until the early stages of incubation

In winter Herring Gulls move about as individuals, either joining a flock or just as readily leaving it. But when breeding begins their behavior changes, and you will start to see pairs of birds flying about together, sharing in all of their daily activities.

Most of the adult birds that have bred in previous years pair up a few weeks before they move to the breeding grounds. When they arrive they reoccupy their territory from previous years and soon start nesting activities.

Young birds that are single go through pair formation on the breeding ground. These birds often gather in "clubs," a term given to groups that preen and rest in an area at the edge of the colony. A male in the club may defend a small space about himself as he gives the Long-call to females flying overhead. The male's call often stimulates the female to land near him.

Following this, the two birds do a series of mutual displays, usually oriented side by side. These usually include the Mew-call, Choking, and finally Head-flagging. After this, one of the birds usually flies off from the spot, and if this is the female, she may return to the same male or land next to another male who is giving the Long-call. After the mutual displays, the male generally becomes aggressive toward other birds in the immediate vicinity.

Birds are considered paired when they begin to stay together throughout the day. This close association is easily recognized and continues up until the time of egg-laying. These paired birds generally relate to each other with new sequences of displays. In one sequence the female approaches the male with Head-tossing, and the male responds either by regurgitating food for the female or by giving the Mew-call and walking with the female to a nearby spot where they do mutual Choking and possibly some scraping of the ground with their feet.

Just before egg-laying, mating becomes much more frequent. One of the birds approaches its mate and begins Head-tossing. If the other bird also responds with Head-tossing, both birds continue with this display while the male aligns himself behind the female. To copulate, the male jumps onto the back of the female and presses his tail down to make sexual contact.

Nest-Building

Placement: On the ground in open areas
Size: Inside diameter 8–10 inches
Materials: Grasses, seaweed, sticks, shells, feathers

After birds have paired they generally move away from the club and occupy a territory. They use the Choking display to assert their presence on the territory and often accompany this display by scraping their feet. These scrapes are usually done near some tuft of grass or small plant and are the first stage in building a nest. Soon one scraping area begins to be used more than the others, and the birds start to gather materials nearby to construct the nest. Both the male and the female help in nest-building. After there are enough materials at the site, one of the birds will sit in the nest, scrape with its feet while turning about, and put material along the rim of the nest with sideways motions of its head.

Locating the Nest

WHERE TO LOOK By the shore or on inland lakes, in gravelly or sandy areas
WHEN TO LOOK From late spring into early summer
BEHAVIORAL CLUES TO NEST LOCATION Breeding colonies are easy to spot, for the Gulls are fairly equally spaced about the area and they give the Alarm-call when you approach.

Breeding

Eggs: 3
Incubation: 26 days, by male and female
Nestling phase: A few days
Fledgling phase: 5 weeks
Broods: 1

Egg-Laying and Incubation

The female may sit on the nest as if she were incubating long before the first egg is even laid, but at this time the pair are still doing all activities together. As soon as the first egg is laid, however, one bird will always stay at the nest to guard the eggs while the other goes to feed. Herring Gulls lay only three eggs in a normal clutch, and there can be a pause of up to three days after the laying of each egg. Incubation does not begin in earnest until the last egg is laid. Before that the birds may spend up to three hours off the eggs just standing by the nest and watching for predators.

The most common predators on Gull eggs are other Gulls. These birds fly over the nesting grounds and swoop in to peck at an egg. They are usually chased away by one of the parents before they can eat any of the egg, but often the egg gets cracked and the developing chick dies.

The parents relieve each other at the nest every few hours. If the brooding bird has been on the nest only for the minimal brood time (about two hours), then the other parent may have trouble getting it off the nest. This may result in some actual pushing and shoving over the nest.

Nestling Phase

There is practically no nestling phase in Herring Gulls. The young hatch over a period of about three days and are brooded for most of the first few days. But when not being brooded, they may wander outside the nest, even on the first day after hatching. When the adults give the Alarm-call, the young tend to scatter and crouch down in protected spots.

Fledgling Phase

For the last week of incubation and the first few weeks after the eggs have hatched, the parents are at their height of aggressiveness toward intruders. There are various degrees of Alarm-calls that you will hear: the lowest intensity is a short, quiet series of *gaga* calls; medium intensity is a louder and longer series of *gagagaga*; highest intensity is signaled by a call of *keew* followed by *gagagaga*s.

The young have two calls: a light begging call and a quavering alarm call. The young react to the Alarm-calls of the adults in the first few days by becoming silent and crouching in the nest. Later, when they can walk freely about the territory, they respond to the Alarm-call by running to cover and staying still and quiet.

When the young beg for food from the parents, they do a display similar to Head-tossing. The parent, in turn, regurgitates food into its mouth and feeds it to the young. Both parents feed the young. This stage lasts for up to five weeks, and at this point the young are as large as the parents. They can fly, but still remain for a short while on the territory. The parents become increasingly aggressive toward the young birds and soon stop feeding them. The young then leave the colony and feed and live on their own.

Plumage

Adult Herring Gulls go through two molts per year. There is a complete molt during August and September, and then a partial molt in March and April of only the body feathers. Young Herring Gulls have plumage marked with various degrees of brown up until their third year, when they appear similar to the adults.

There is no way to distinguish the sexes through differences in their plumage, but when a pair is together, the larger bird is usually the male. The male is also the main member of the pair to be involved with aggressive interactions.

Seasonal Movement

Herring Gulls need open water in winter and a fairly secluded breeding ground in summer. The birds are known to migrate north in spring when lakes and rivers become free from ice. In late spring they may then make a shorter movement to islands along the seacoast or to secluded spots along lakeshores to begin breeding. Many gulls do not breed, and they remain in flocks that often feed at city dumps during the day. In late summer, the birds leave their breeding grounds and gather along large rivers or along the coast, and when the freezing weather sets in they usually migrate south.

Pigeon / *Columba livia*

I HIGHLY RECOMMEND PIGEONS FOR YOUR FIRST EXPERIENCE OF observing visual displays. With just a few minutes of watching any large active flock, you are bound to see several exciting displays such as Bowing, Tail-drag, Driving, and Wing-clapping-flight. The best places to find these flocks are city parks, where the birds are accustomed to humans and usually gather to feed. To see the displays, first familiarize yourself with them by looking over the display guide, then go out, have a seat on a park bench, and scan the flocks for any unusual movements. The displays seen on the feeding ground are primarily those associated with courtship.

To see other displays, go to an area where the birds are nesting. Pigeons prefer to nest where there are dark cubbyholes fairly high above the ground, such as on churches, under bridges, or on the facades of elaborate buildings. Here you are apt to see a whole new set of displays (Billing, Nodding, and mutual preening) as well as many of the displays seen on the feeding ground.

You can observe much of the behavior of Pigeons without really trying. Nest-building is one such pattern. At some point you are bound to see a Pigeon pick up a short, straight twig in its bill, shake it, and then fly off with it. This is most likely the male, for he collects most of the material for the nest and has the habit of shaking the collected material before taking it back to the nest.

Another behavior pattern that may catch your eye is Billing.

This is where the male opens his beak and the female places her beak inside his, and in this position the two birds bob their heads up and down. This is a form of mate-feeding where the female actually receives food regurgitated from the male's crop. The young fledgling receives food in this same way from either parent until it is independent.

The Pigeon breeding cycle generally starts in spring, and continues through summer, but in warmer parts of the continent you may see certain aspects of breeding in fall and winter as well.

BEHAVIOR CALENDAR

	TERRITORY	COURTSHIP	NEST-BUILDING	BREEDING	PLUMAGE (MOLTS)	SEASONAL MOVEMENT	SOCIAL BEHAVIOR
JANUARY		■					
FEBRUARY		■	■				
MARCH	■	■	■	■			
APRIL	■	■	■	■			
MAY	■	■	■	■			
JUNE	■	■	■	■			
JULY	■	■	■	■			
AUGUST				■	■		
SEPTEMBER					■		
OCTOBER							
NOVEMBER		■					
DECEMBER		■					

DISPLAY GUIDE

Visual Displays

Bow

Male or Female *Sp Su F W*

With neck feathers ruffled, the bird lowers its head and turns in full or half circles.

CONTEXT Done primarily by males: after landing near a flock, during assertion of territory, or in front of a prospective mate. *See* Territory, Courtship

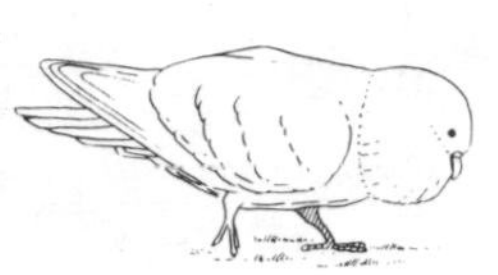

Tail-Drag

Male *Sp Su F*

Between repeated Bowing, the bird lifts its head high and runs for short distances, dragging its spread tail along the ground.

CALL Tail-drag-call

CONTEXT Occurs most on the feeding ground and is done by males in front of their mates. *See* Courtship

Wing-Clapping-Flight

Male or Female *Sp Su F W*

As a bird takes flight, its wings hit together making a clapping sound. In flight the bird may alternate Wing-clapping and gliding. In the glide its wings are held in a V and its tail is spread.

CONTEXT Occurs in many situations but usually is done by only sexually active birds. *See* Courtship

Driving

Male and Female *Sp Su*

Two birds run in tandem away from the flock or in among it. The second bird seems to drive the first along.

CONTEXT Occurs between members of a pair when other birds are present, with female preceding male. *See* Courtship

Billing

Male and Female *Sp Su*

A female puts her bill into her mate's open mouth, and the two move their heads rhythmically up and down together.

CONTEXT Occurs between members of a pair often just prior to mating. *See* Courtship

Nodding

Male and Female *Sp Su F W*

While crouched down, the bird repeatedly nods its head, as if it were slowly pecking.

CALL Nodding-call

CONTEXT Occurs at possible nest sites. *See* Nest-building

Allo-Preening

Male and Female *Sp Su W*

Two birds do this together, one pecking lightly at the head feathers of the other.

CONTEXT Occurs between members of a pair, at or near a nest site. *See* Courtship

Displacement-Preening

Male or Female *Sp Su*

A bird quickly touches behind its wing with its bill.

CONTEXT Often given prior to or in response to other displays. *See* Courtship.

Auditory Displays

Wing-Clap

Male or Female *Sp Su F W*

A clapping sound produced in Wing-clapping-flight

CONTEXT Usually given by sexually active birds in a variety of circumstances. *See* Courtship

Nodding-Call

Male or Female *Sp Su F W*

A long, drawn-out *coo* that can be heard quite easily *k't'coooo.*

CONTEXT Given with the Nodding display. *See* Nest-building

Tail-Drag-Call

Male *W Sp Su*

A variable series of quiet *coos*; not heard unless given close by *oorook'toocooo.*

CONTEXT Given with Tail-drag display. *See* Courtship

BEHAVIOR DESCRIPTIONS

Territory

Type: Nesting
Size: Immediate area of nest site
Main behavior: Bow, striking with outstretched wing
Duration of defense: Throughout the breeding cycle

Territory is a relatively minor aspect of Pigeon life. Usually only the immediate area around the nest site is defended against intruders. Occasionally, when the bird is nesting at the edge of a porch roof or similar structure, it may extend its territory to the adjacent roof area. The Bow, given by either the male or the female, is usually the first display directed toward an intruder. More vigorous defense involves the tail's being spread, plumage ruffled, bill gaped, and one wing outstretched and lifted. If actual fighting occurs, the intruding bird is struck with the lifted wing.

Courtship

Main behavior: Bow, Tail-drag, Driving, Displacement-preening, Billing, Wing-clapping-flight, Allo-preening
Duration: Extremely variable

Probably the most entertaining aspect of Pigeons' behavior is the variety and frequency of their courtship displays. They have at least six conspicuous courtship displays that can be seen throughout most of the year.

The displays can be divided roughly into three groups: early displays, which occur mostly on the feeding ground; mating displays, which more often occur near the nest site; and nest site demonstration displays, which occur at the nest site. The first two are described here; the third is included in the Nest-building section.

The first phases of courtship can be seen best in large groups of Pigeons that have gathered to feed. Birds leaving or entering the flock will often do Wing-clapping while in flight. This display is done most often by males, and it may function to advertise sexual maturity. It generally has the effect of stimulating other birds, males and females, also to take flight with Wing-clapping. Males that have recently landed near a feeding group may do Bow or Tail-drag in front of a female in the flock. The female may appear to ignore this display, but often she increases the speed of her pecking the ground without actually picking up food. A fourth display associated with these others is Driving, in which the male seems to chase the female and at times even pecks at her. This set of displays is characteristic of the period one to two weeks before the first egg is laid.

In the last week before egg-laying, nesting and mating displays become more common. Mating displays include Billing and Displacement-preening. Billing is the more obvious, for in it the female puts her beak into the male's open beak and the two bob their heads up and down. This is often preceded by the more subtle Displacement-preening, where the bird quickly places its bill up behind its wing for only a moment. This movement does not function as preening, but as a signal of a certain state of readiness in the displaying bird. Both these displays are closely associated with copulation, either preceding or following it.

Another behavior pattern seen at the nest site is one bird's preening its mate on the head and bill region (Allo-preening). Between one and two weeks before nesting, the female usually preens the male. As nesting starts, this reverses, and the male most often preens the female.

Pairs of Pigeons are believed to mate in consecutive seasons for as long as both birds of a pair live. This is most likely due to their sedentary habits and their almost continual reproductive readiness. The length of courtship among Pigeons is extremely variable, depending to a great extent upon the maturity of the birds and whether they have previously raised broods together.

Nest-Building

Placement: On building ledges, bridge supports, or similar natural sites
Size: 8 inches outside diameter
Materials: Short stiff twigs or other material with similar qualities

Nest-building in Pigeons involves a great deal of interaction between the pair. The male first chooses a nest site, and then, in view of the female, he crouches down and repeatedly Nods toward the ground. In time the female joins him at the spot, crouches beside him, and starts to join him in Nodding. In a short while they may begin to quiver their wings nearest each other and then add cooing calls with each nod. Following this the male will go in search of nesting materials while the female stays at the site.

Pigeons make their nests out of a collection of stiff twigs, the stiffer the better. You can often see a male picking up twigs, adjusting them in his bill, shaking them and then dropping them. This seems to be a way of testing the twigs for stiffness. Having found a suitable twig (only one at a time) he carries it to the nest site and lays it in front of the female. She then picks it up and places it into the nest underneath her. This process continues until the nest is built.

Locating the Nest

WHERE TO LOOK Along building ledges and ornaments and on the supports of bridges in country or urban areas

WHEN TO LOOK The months from February to August; winter months also in the warmer areas of the country

BEHAVIORAL CLUES TO NEST LOCATION:

1. There may be an accumulation of droppings on the ground underneath possible nest locations.
2. Watch for males, and more rarely females, gathering twigs and follow the birds' flight as they return to the nest.

Breeding

Eggs: 2
Incubation: 18 days, by male and female
Nestling phase: 10 days
Fledgling phase: Little or none
Broods: 2-3

Egg-Laying and Incubation

The female may sit for long periods on the nest a day or two before the first egg is laid. Once the first egg is laid the nest is rarely left unattended. Two eggs is the normal clutch size. Both male and female incubate the eggs, but generally only the female remains on the nest at night — the male roosts away from the nest, even though he may have roosted with the female before the eggs were laid. Incubation lasts for seventeen to eighteen days.

Nestling Phase

At first, the parents feed the young with regurgitated food from their crops. This has been called "pigeon milk" but it is not related to the milk of mammals. Over the period of ten days in the nest, the young are fed increasing amounts of solid food such as caterpillars, fruits, etc.

Fledgling Phase

Young Pigeons out of the nest will continue to beg for food from their parents, but within a few days they are on their own. If they return to the nest site while the parents are engaged in raising a second brood they will be forcibly ejected from the nest area, especially by the female parent.

Plumage

Pigeons go through one complete molt per year. This occurs in late August and early September. You cannot distinguish male Pigeons from female Pigeons through their appearance, but you can tell the difference through their behavior. Although Bowing is done by both sexes, the majority of the time it is done by the male. The male is also the only one to do Tail-drag. In general the male is the one to collect the nest material, while the female is the one who constructs the nest.

Seasonal Movement

Pigeons have no seasonal movement patterns. They stay in roughly the same areas throughout the year.

Feeder Behavior

In city and suburban areas, Pigeons may come to home bird feeders if food is placed on the ground. But an even more likely spot to see them feeding is in a city park, where they glean bits of food intentionally or unintentionally left on the ground. Much of their courtship behavior can be seen at these times. See the section on Courtship.

Most common displays: During feeding look for the Bow, Tail-drag, and Driving displays. When a new bird flies into a group of other Pigeons, it often does Bowing just after landing. Wing-clapping-flight may also occur as the birds arrive at or leave the feeding site. See the Display Guide.

Other behavior: The Tail-drag-call is soft, but if you are close, it may be heard from a bird doing the Tail-drag.

Chimney Swift / *Chaetura pelagica*

THE ARRIVAL OF SWIFTS IN LATE SPRING IS AN EXCITING MOMENT. You will probably first hear their chittering calls, then look up and see their small bodies and curved, bladelike wings slicing through the air in graceful arcs. Their constant flight throughout the day makes them both physically and experientially remote from our earthbound living. But even so, bending back your neck to see only sky and these streamlined birds gliding about can draw you into their world of flight — so much so that when you again look down it may take a moment to reorient yourself to the world of the flightless.

More than any of our other common birds, the Swift's life is lived on the wing. As migration gets under way, large flocks can be seen in the early evening flying in formation over possible roosting sites and calling loudly all the time. Then, as it starts to get dark, they begin to dive into the roost, sometimes forming a steady stream out of the airborne flock.

Courtship among Swifts seems to take place primarily in flight. The first behavior pattern you will see is birds' remaining together in small groups as they circle high in the air. Slightly later groups of just three birds may be seen. During these flights there are loud chittering calls and two of the birds flying close together seem to follow a third that is farther ahead. The most common flight display involves just two birds flying together — one slightly above the other. The hind one abruptly lifts its wings into a V and glides — the one in front then does the same. This display can be seen all summer.

When the pair have selected a nest site they begin to gather short straight twigs for the nest, grabbing them off trees with their feet as they fly by. The nest is usually built on a vertical surface of wood or stone and is cemented to this surface with the birds' saliva.

Both parents share in the incubation of the eggs, and about four weeks after hatching, the young take their first extended flights. The Swifts begin their migration in late August, some traveling as far as the rain forests of Peru, where they spend the winter before returning the next year.

BEHAVIOR CALENDAR

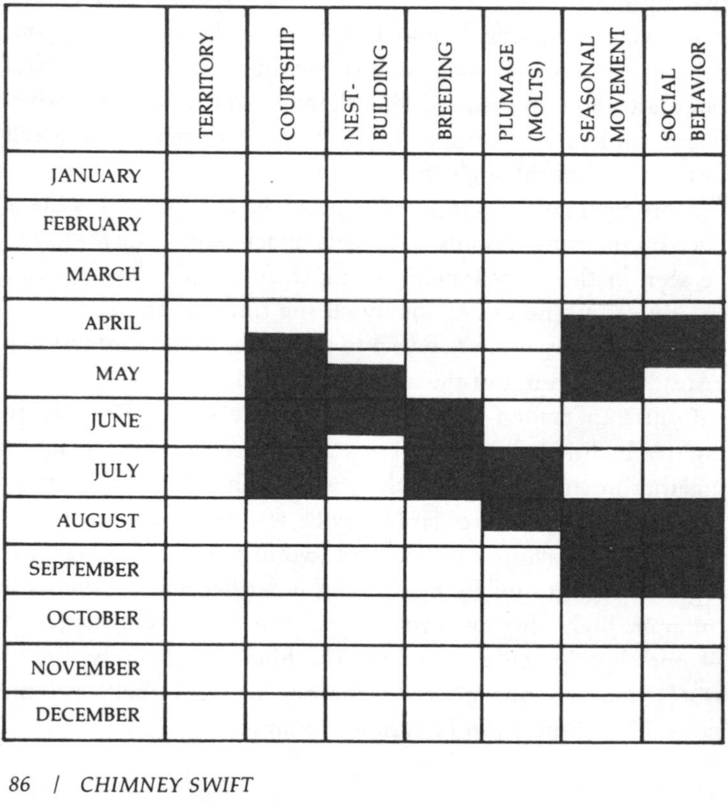

	TERRITORY	COURTSHIP	NEST-BUILDING	BREEDING	PLUMAGE (MOLTS)	SEASONAL MOVEMENT	SOCIAL BEHAVIOR
JANUARY							
FEBRUARY							
MARCH							
APRIL							
MAY							
JUNE							
JULY							
AUGUST							
SEPTEMBER							
OCTOBER							
NOVEMBER							
DECEMBER							

DISPLAY GUIDE

Visual Displays

Trio-Flight

Male and Female *Sp*

Three birds fly in follow-the-leader fashion for up to five minutes, the last two birds closest together.

CALL Chip-call

CONTEXT Seems to contain two males following a female; occurs prior to breeding. *See* Courtship

V-Glide

Male or Female *Sp Su*

One or both members of a pair, flying in close tandem, snap their wings into a rigid V above their backs and glide.

CALL Chitter-call

CONTEXT Seems to occur between members of a pair and can be seen all through the breeding cycle. *See* Courtship

Auditory Displays

Chitter-Call

Male or Female *Sp Su*

A rapidly repeated, high-pitched call *chitterchitterchitter*

CONTEXT Often given while the birds are in flight, and in a variety of situations

Chip-Call

Male or Female *Sp Su*

A quick series of separate *chips* *chip, chip, chip, chip,*

CONTEXT Given during Trio-flights. *See* Courtship

BEHAVIOR DESCRIPTIONS

Territory

There are no reports of territorial behavior in the Chimney Swift, although it is likely that it defends the nest site.

Courtship

Main behavior: Group flights, Trio-flights, V-gliding
Duration: Variable

The life of the Chimney Swift is one of the hardest to study, especially in those aspects that take place while the bird is in flight. Courtship is one of these activities, and although we cannot yet point to any one display and say that it is a courtship display, there are some behavior patterns that may serve this function. They include three types of flight.

The first type starts within two weeks of the birds' return to their breeding grounds. It has been called "together flight" and involves groups of four to seven birds flying together in a circular or oval pattern. The birds stay in a loose group, some occasionally leaving for a moment and then rejoining the group. The birds are not feeding, as can be seen by their lack of erratic movements to catch insects.

Another flight pattern that occurs slightly later is Trio-flight. This flight can be recognized by the fact that it contains just three birds and by the predictable spacing of the birds. Typically there is one leader bird and two birds behind. The two following birds are closer to each other than either of them is to the leading bird. Trio-flight may be low, with the birds darting in and out among the treetops, or it may ascend high into the air. At least one of the three birds gives the Chip-call during the display. Close observation of this formation by one researcher shows a tendency for the order of the birds to stay the same, and for the two following birds to seem to compete for second place in the formation.

Another common behavior of Chimney Swifts that may be associated with pair formation is V-gliding. This involves two birds flying in tandem, the second one slightly above the first and very close to it. Suddenly the second bird will snap its wings into a V above its back and glide. In many cases the front bird will then do the same. After gliding for five to ten seconds the birds resume normal flight. The Chitter-call may accompany the display.

The V-gliding display may be done during Trio-flights or it may be done by just two birds. Early in the breeding season just one member of a pair may hold its wings in a V, but later in the season both do it together. The display can be seen all summer and even into autumn.

Mating among Chimney Swifts takes place at the nest site, and not in the air as has been believed in the past.

Nest-Building

Placement: In dark, protected areas of buildings, especially chimneys
Size: 4 inches in diameter
Materials: Short straight twigs and the bird's saliva

Although few people will have the chance to watch nest construction by Chimney Swifts, since their nests are located in inaccessible spots, anybody, with a little luck, can watch their interesting behavior as they gather the nest material. A Chimney Swift gathers nest material while in flight, first choosing a tree with many dead twigs of the desired dimensions and then flying in a circle about it. Finally it dives at the tree and grabs a twig with its feet, trying to break it off as it flies by. If it fails to do so it will fly up and come back at the twig again. As the bird flies back to the nest it transfers the twig from its feet to its bill. This activity is best seen in the afternoon and evening.

Both the male and the female participate in building the nest, which may occur over a period of up to thirty days. The pieces of twig that are brought to the nest are cemented onto a vertical surface with a secretion from the bird's mouth, often termed "saliva" although it doesn't have the same function as our saliva. When the nest is about half finished, the eggs are laid and incubation begins. The rest of the nest, consisting of supports fastened at the upper edge of the nest and to the wall above the nest, is finished during the incubation period. Old nests or nest sites are often reused, but only after some repairs.

Locating the Nest

WHERE TO LOOK In chimneys or around the eaves of old barns, or buildings where the birds can easily get in and out; generally high above the ground

WHEN TO LOOK From May into June, in the afternoon or evening

BEHAVIORAL CLUES TO NEST LOCATION:

1. Watch for the birds circling quietly over trees and then dropping down into them; or watch for the birds carrying twigs in their beaks and follow where they go.
2. Later in the season the birds will be making trips to the nest with food, so be on the lookout for any Swift that enters a building.

Breeding

Eggs: 4–5
Incubation: 19 days, by male and female
Nestling phase: 14–19 days
Fledgling phase: 14–18 days
Broods: 1

Egg-Laying and Incubation

The eggs are laid one every two days until the clutch is complete. Four to five eggs is the average clutch size, and the eggs are incubated by both the male and the female. Incubation is usually nineteen days but can take up to twenty-one days.

Nestling Phase

In the first week of nestling life, the parents alternate brooding the young. When a parent comes to the nest to feed the young, the other parent leaves to gather food. After feeding the young, the parent that just brought food takes a turn at brooding. In this way they alternate food searching and brooding. At first the young are fed regurgitated insects, but later they are fed a mass of insects collected by the parent and held together like a pellet by a sticky substance from the parent's mouth. The young stay in the nest for fourteen to nineteen days.

Fledgling Phase

After the nestling phase the young leave the nest, but they do not go far; rather they remain clinging to the walls in the immediate area of the nest and take practice flights of only a few feet around the nest. In this stage they are still dependent on their parents for all of their food. After two weeks remaining near the nest, the young finally fly out into the open away from the nest. From this time on they feed themselves. They do not yet stay on the wing all day, but return to the nest during the day and roost there at night. After a week of this behavior they are strong enough to fly all day, and they remain on their own.

Plumage

It is believed that the Chimney Swift goes through two molts per year. One occurs in late summer, and this is a complete molt. The other occurs before the spring migration, and it is not yet known whether this is a complete molt or only a molt of the body feathers, leaving the wings and tail intact.

The plumage of both sexes is identical, and there is no way known to distinguish the sexes by behavior.

Seasonal Movement

Among our common birds the Swifts remain for the shortest time on their breeding ground before flying south for the winter. In spring they enter North America along the Gulf Coast in mid-March, and they are on their way south again during August and September. They fly north in small flocks of twenty to thirty birds, but fly south in fall in large flocks of hundreds of birds. See Social Behavior for descriptions of their interesting roosting habits during migration.

Social Behavior

During spring and fall migrations Chimney Swifts roost communally. It is exciting to watch the birds flock together and then enter these roosts. Up to an hour or more before sunset the birds may begin to gather, giving their Chitter-calls. Often they arrive at the roost in a long, drawn-out formation, but once there, circle around to form an O. From minute to minute this formation may change, some birds breaking away from the group and all following them and then returning again to the circular formation. This may continue for over forty-five minutes before the birds start to enter the roost. Out of the swarming circle a few birds may drop down to the roost spot and then fly back up to rejoin the group. Soon a few birds will enter the roost site and will be followed by a continuous thread of birds behind them until all are in the roost. Even when they are in the roost their Chitter-calls can be heard for another fifteen minutes or so, and then the group quiets down.

Roost sites are generally steeples or old chimneys. The birds leave in the morning suddenly and seemingly without ceremony — a few minutes after they have started to leave, no Swifts will be seen in the area. Their time of leaving is often quite irregular. These roosting groups can be seen mainly during spring migration and just before breeding begins, and then again in fall as the birds gather into large flocks to start their migration to the south.

FL

Common Flicker / *Colaptes auratus*

THE COMMON FLICKER IS A CONSPICUOUS BIRD BECAUSE OF ITS loud calls and its active behavior. But since a single bird may range over an area of 150 acres or more, it is almost impossible to follow all the aspects of any one individual's life. You are more likely to come across the birds by circulating through areas where you frequently hear their calls, and thus pick up on bits and pieces of their behavior.

The area around the nest often has the most activity. The nest holes are not too hard to discover, as the birds return to the same area each year and often to the same tree to excavate a new nest. Finding fresh wood chips on the ground may lead you to the discovery of a Woodpecker excavation.

The most active displaying occurs early in the season, before nest-building, when the birds are pairing and there is competition for mates. Head-bobbing is the most common visual display of Flickers, and it is accompanied by the conspicuous *woikawoikawoika* call. It occurs most in spring and then is seen again in fall just before the birds migrate south. When it is done between a male and a female, it concerns courtship; when it is done between two birds of the same sex, it is generally competition for a mate or, more rarely, is a territorial dispute. Thus it is important to be able to tell male from female to understand the nature of the display. This is easily done, for the male has black streaks extending from the back of his bill across either cheek, like a mustache, and the female has no streaks.

The auditory displays of the Common Flicker are few and

easily distinguished. They offer an interesting challenge to the behavior-watcher, for their functions are still generally unclear.

Drumming, which is such a prominent part of the behavior of other common Woodpeckers, is less so with the Flicker. Its place seems to be taken by the loud Kekeke-call, which can be heard for long intervals at the beginning of the breeding season.

BEHAVIOR CALENDAR

	TERRITORY	COURTSHIP	NEST-BUILDING	BREEDING	PLUMAGE (MOLTS)	SEASONAL MOVEMENT	SOCIAL BEHAVIOR
JANUARY							
FEBRUARY							
MARCH							
APRIL							
MAY							
JUNE							
JULY							
AUGUST							
SEPTEMBER							
OCTOBER							
NOVEMBER							
DECEMBER							

DISPLAY GUIDE

Visual Displays

Head-Bobbing

Male or Female *Sp Su F*

The bird raises its breast and bobs its head up and down and from side to side. The wings may be lifted slightly to expose the colorful underwings, and the tail may be spread and tilted sideways to reveal its colorful underside.

CALL Head-bobbing-call

CONTEXT Used primarily in spring during intrasexual rivalries for mates, during the first stages of courtship, and during territorial conflicts. *See* Territory, Courtship

Auditory Displays

Drumming

Male or Female *Sp*

A rapid hitting of a resonant surface with the tip of the bird's bill; like a drumroll

tatatatat

CONTEXT Occurs mostly around the nest tree and has the effect of attracting nearby Flickers to the spot; often given between mates. *See* Territory, Courtship

Kekeke-Call

Male or Female *Sp Su F*

A repeated, loud, percussive sound easily heard from a great distance

kekekekekeke

CONTEXT Heard frequently upon first arrival of the birds in spring and again about a month later when courtship activities are renewed; given mostly between mates. *See* Territory, Courtship

Keeogh-Call

Male or Female *Sp Su F*

keeogh. keeogh. A loud, slightly drawn-out sound, given singly

CONTEXT Used between members of a family to help them locate each other

Head-Bobbing-Call

Male or Female *Sp Su F*

woikawoikawoika A repeated, two-syllable phrase that sounds much like the written description

CONTEXT Accompanies the Head-bobbing display. *See* Territory, Courtship

Quiet-Notes

Male or Female *Sp Su*

weetaweetaweeta Notes with a quality similar to the Head-bobbing-call, but softer and quieter, and occurring without Head-bobbing display

CONTEXT Occurs between mates whenever they come together, especially near the nest site. *See* Courtship

Warble-Note

Male or Female *Sp Su*

A short, low-pitched, throaty warble

CONTEXT Given especially near the nest when there is some possibility of danger, such as your own presence

BEHAVIOR DESCRIPTIONS

Territory

Type: Mating, breeding, feeding
Size: ½ mile square range in vicinity of nest; territory smaller
Main behavior: Head-bobbing, frozen pose, Kekeke-call, chases, Drumming
Duration of defense: From arrival on the breeding ground through the fledgling phase of the last brood

Flickers are fairly consistent in returning to their breeding ground year after year. In fact, when first arriving in spring, they often return to the area of their previous nest tree. From these spots the males announce their presence with loud volleys of the Kekeke-call. They also give brief volleys of Drumming from their nest tree or nearby drumming posts. Territorial interactions occur when a new bird flies into the area. They happen early in the season and involve just two males or two females; interactions of three or more birds involving both sexes are described under Courtship. In the two-bird interactions, the Head-bobbing display with its accompanying call is common. Interspersed with the Head-bobbing will be moments of the frozen pose where both birds remain absolutely still, then resume Head-bobbing or a chase. Check for the sexes of the birds (*see* Plumage), for courtship interactions are similar, but take place between a male and a female.

The range of a Flicker is an area of about one-half mile square, but the consistently defended area is much smaller and is composed of the area around the nest site. Before egg-laying, territorial conflicts may occur more generally about the range, but after egg-laying, they are limited to the vicinity of the nest.

Courtship

Main behavior: Head-bobbing, Drumming, Kekeke-call, Quiet-notes, triangle encounters
Duration: For migrants, about 2–3 weeks; for residents, as long as 2 months

Flickers usually mate for life, since both birds return to the same area to breed year after year. Each bird announces its arrival on the breeding ground with Drumming or the Kekeke-call. These tend to bring the members of a pair together at the same spot, whereupon they may greet each other with vigorous Head-bobbing displays as well as with the frozen pose. The displays are the same as those in territorial skirmishes, the only difference being that these involve one male and one female displaying to each other — the others involve members of the same sex. As the male and female continue to meet during the courtship period, their Head-bobbing displays become less intense and the accompanying calls get softer and are finally replaced by the Quiet-notes. Soon the Head-bobbing display is no longer used between the pair, but whenever they come close together they still both utter the Quiet-notes.

Early in the breeding season another type of interaction is commonly seen, for it is noisy and active. This behavior involves three birds, two of one sex and one of the other. Generally it occurs when a new bird enters an area and competes with one member of a pair for its mate. Typically, the third member, being competed over, remains nearby but out of the way, often just feeding. The other two competing birds use Head-bobbing, frozen pose, and chasing to establish which one is to be dominant. In most cases the intruder is forced to leave.

Courtship activities are renewed for a short period about one week before the nestlings fledge. At this time frequent Kekeke-calls will be heard and the pair may interact frequently. Courtship activities also seem to take place in fall before the birds migrate from the north. These are only brief and not in relation to territory or a nest hole — they may be interactions between first-year birds.

Nest-Building

Placement: Between 10 and 30 feet off the ground in a dead wood stump
Size: Round hole about 2 inches in diameter
Materials: Excavated tree hole, lined with only wood chips

Flickers have slightly curved bills and are weak excavators. Therefore they generally excavate dead wood that is well weathered or partially rotted. Since there are not always a number of trees in the area that meet these requirements, the Flickers often excavate new holes in the same tree year after year. These trees are obvious when you come across them and are good places to look for future nests.

Both male and female participate in excavation, but the male does the majority of the work. The birds excavate only for short periods, and the times of day when they excavate are irregular. Many nest holes are started but never completed, possibly due to wood quality or nest location. Sometimes the birds will reuse an old hole after doing some minor work on it.

Locating the Nest

WHERE TO LOOK Anywhere where old, large trees exist that have some dead or rotting wood on them well above the ground, especially near open areas, where the birds feed

WHEN TO LOOK In April and May, when excavation starts

BEHAVIORAL CLUES TO NEST LOCATION:

1. Listen for any Flicker drumming, for it usually does this very near the nest site.
2. Look for fresh wood chips at the base of trees with dead wood.
3. Listen for the Warble-note; it is often given as a bird leaves the nest.

Breeding

Eggs: 7–9
Incubation: 11–12 days, by male and female
Nestling phase: 26 days
Fledgling phase: 2–3 weeks
Broods: 1–2

Egg-Laying and Incubation

Once the egg-laying period has started the nest is always attended by at least one of the adults. Certain perches near the nest become important resting and preening spots. The male and female share in incubation, taking turns that average about an hour each. The male, however, always incubates the young at night while the female roosts in a hole nearby. The birds are generally quiet during this phase of the breeding cycle.

Nestling Phase

The day before the eggs hatch, the parents exchange places on the nest more frequently than before. The young may hatch over a two-day period, and for the first few days they are brooded at night, again by the male only. Unlike other common Woodpeckers, the Flicker feeds its young through regurgitation. Early-morning feedings can be as frequent as once every ten minutes, but later in the day there may be as few as one per hour. After the third week the young are strong enough to crawl up to the nest hole and be fed. To see the parent feeding the young at the entrance, then, means that the young will fledge within a week.

Fledgling Phase

The young are strong fliers upon leaving the nest, and in many cases stay with the parents for two to three weeks. At this time the young may still be fed by the adults, and the whole family group may move far from the nest but remain on the range. When the young are independent they leave the family group.

Plumage

Adult Flickers go through one complete molt per year. This occurs from late July through September.

Male and female Flickers can easily be distinguished on the basis of their plumage, for the male has two black streaks on either cheek, often referred to as a mustache, while the female's cheeks are unmarked. The only exception to this is that the juveniles of both sexes have black cheek marks until their first fall molt when they become like the adults.

Seasonal Movement

Flickers from northern areas migrate south in fall, usually moving in small, loose flocks that at times may grow to include as many as one hundred birds. Before migrating, Flickers will be seen in small, active groups of from five to ten birds. These are conspicuous and individual interactions are lively, often involving the Head-bobbing display.

In southern areas Flickers will be present throughout the year, but in winter their population will increase slightly due to northern migrants. The migration north in spring is usually started by the males, and the females follow a few days later.

Hairy Woodpecker / *Dendrocopos villosus*

FROM LATE WINTER TO EARLY SUMMER YOU CAN HEAR WOODpeckers Drumming. This is the loud, continuous pecking on hard surfaces that they use as a signal to announce territory and attract mates. Drumming serves roughly the same function as song does in many other birds, and once you start hearing it, you know that the Woodpeckers' breeding season has begun. I once heard a Hairy Woodpecker Drumming and decided to track it down. I soon found it and stayed quietly nearby to watch its behavior. The bird repeatedly Drummed on the same spot, and each time it stopped, it looked around in all directions. After about fifteen minutes it began to move about and do short volleys of Drumming on other parts of the tree. In each new spot, the Drumming had a different quality and loudness, depending on whether the wood was dry, rotted, or hollow underneath, but none of the new spots were as loud as the first. After a few more minutes it returned to the original spot and continued in the same manner that I had first seen.

Drumming is one of the more exciting aspects of Woodpecker behavior, for it alerts you to an important spot in a Woodpecker's territory. In areas where there is Drumming, there is apt to be other interesting behavior, such as sexual interactions. Early in the season these take the form of triangle disputes where one member of a pair is being challenged for its mate. These conflicts may be quite prolonged and involve flight displays alternated with periods of Still-pose where the birds remain within sight of each other but absolutely motionless for up to a few minutes at a time.

Later in spring you can watch the excavation of the nest hole. Pecking for feeding or excavation consists of light taps in irregular rhythms, and has an entirely different sound from Drumming. Behavior-watching in the area of the nest is always filled with the possibility of seeing interesting interactions between the pair. Once the young have hatched, their calls can be heard from quite some distance as they continually *chirp* from inside the tree hole. Again, as the parents come and go bringing food, it is a good chance to see how they coordinate their activities.

BEHAVIOR CALENDAR

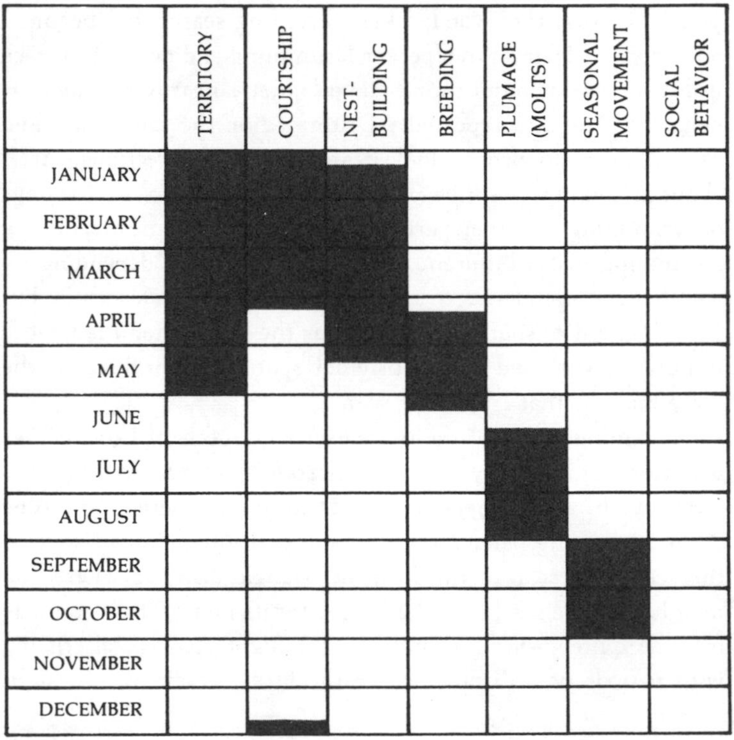

	TERRITORY	COURTSHIP	NEST-BUILDING	BREEDING	PLUMAGE (MOLTS)	SEASONAL MOVEMENT	SOCIAL BEHAVIOR
JANUARY							
FEBRUARY							
MARCH							
APRIL							
MAY							
JUNE							
JULY							
AUGUST							
SEPTEMBER							
OCTOBER							
NOVEMBER							
DECEMBER							

DISPLAY GUIDE

Visual Displays

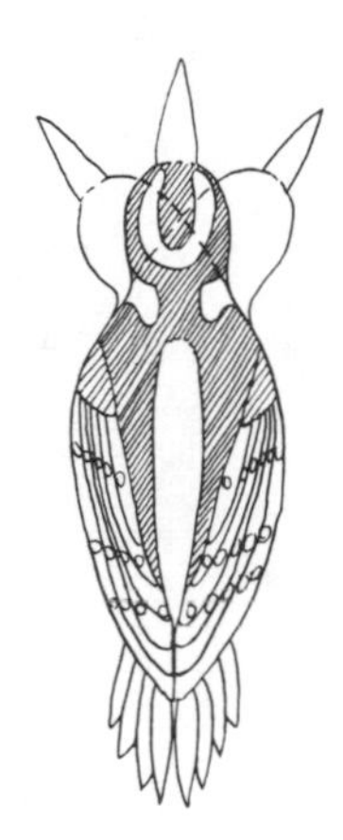

Bill-Waving

Male or Female *W Sp Su*

The bird repeatedly waves its bill from left to right. Along with this are other actions: tail spread, wings flicked, breast feathers fluffed, and nape feathers fluffed.

CALL Bill-wave-call

CONTEXT Given during intrasexual conflicts; also given to some extent between members of a pair in courtship. *See* Courtship, Breeding

Fluttering-Flight

Male or Female *W Sp Su*

The bird beats its wings rapidly, almost hovering, with its head held down.

CONTEXT Used by a bird at significant times when it is leaving or approaching its mate; generally considered a part of courtship, but also used by adults when carrying fecal sacs from the nest. *See* Courtship, Breeding

Still-Pose

Male or Female *W Sp Su*

Two birds, after interacting, suddenly stop all movement and remain absolutely still for anywhere from one to twenty minutes.

CONTEXT May happen during courtship displays, but more often is done in the middle of aggressive encounters. *See* Courtship

V-Wing

Male or Female *W Sp Su F*

The bird, either perched or in flight and about to land, raises both wings at a forty-five-degree angle from the body.

CALL Bill-wave-call

CONTEXT Occurs during aggressive encounters with other birds of the same or other species; sometimes given toward nonavian predators when they endanger the nest. *See* Courtship, Breeding

Swoop-Flight

Male or Female *W Sp Su*

The bird makes a deep downward loop in flight just before landing.

CONTEXT Occurs before a bird lands at a point of special interest to it, such as a Drumming post or nest site

Auditory Displays

Bill-Wave-Call

Male or Female *W Sp Su*

eetickiwickiwickiwicki An energetic call very much like its written description

CONTEXT Always given with the Bill-waving display. *See* Courtship

Teek-Call

Male or Female *Sp Su F W*

teeek. teeek. A single sound given at spaced intervals

CONTEXT Given all year, and may be used between members of a pair to keep in contact as well as in other situations. *See* Courtship

Tew-Tew-Call

Male or Female *W Sp Su*

A series of soft notes; a quiet call — *weetew weetew weetew*

CONTEXT Given between members of a pair during the later stages of courtship and throughout breeding whenever they rejoin after a separation. *See* Courtship, Breeding

Khirr-Call

Male or Female *W Sp Su*

A drawn-out call a little like the whinny of a horse — *khirrrrr* or *teek-teek-khirrrrr*

CONTEXT Often given in combination with the Teek-call during moments when a bird is asserting itself on its territory or range. *See* Territory, Breeding

Keke-Call

Male or Female *W Sp Su*

A loud series of single notes closely spaced — *kekekekeke*

CONTEXT Given during aggressive interactions. *See* Courtship

Drumming

Male or Female *W Sp*

A rapid hitting of a resonant surface; faster than one can count — *tatatatatat*

CONTEXT Given in both courtship and territory formation; has a variety of functions. *See* Territory, Courtship

Wing-Clapping

Male or Female *W Sp Su*

A sound like muffled clapping, created by the wings hitting each other (presumably)

CONTEXT Given in connection with flight displays during courtship, aggressive encounters, and in the ritual preceding the changeover at the nest. *See* Courtship, Breeding

BEHAVIOR DESCRIPTIONS

Territory

Type: Nesting
Size: 1/4 acre; range is much larger
Main behavior: Khirr-call, Drumming, perch-taking, evasive action and chases
Duration of defense: From the choice of a nesting tree up to the fledgling phase

With the Hairy Woodpecker it is important to distinguish between range and territory. In general a territory is an area that a bird consistently defends against other birds of its own species, whereas a range is an area the bird stays within but does not defend. Hairy Woodpeckers live on a large range of about six to eight acres throughout the year. Although this range is not consistently defended, the bird does assert itself on the range by the Khirr-call and by Drumming on certain resonant trees often called "signal posts." The ranges of neighboring Woodpeckers may overlap, and no bird is dominant over all other birds on its range, as it is in its territory.

In midwinter, when courtship begins, the birds become slightly more assertive on their ranges and patrol them more often. Actual territorial defense starts later and is only occasionally seen, for it occurs over a long period of time. Also, since the birds are year-round residents, they are familiar with the movements of their neighbors, and so are less likely to get into conflicts with them.

The real territory is small and surrounds just the nest site. It is carefully defended as soon as the birds have chosen a nest tree, even before excavation has started. Aggressive activities on the territory are less conspicuous then those on the range, usually involving just perch-taking or quick attacks upon an intruder to drive it from the area. Sometimes the territorial bird will hitch around a tree out of sight when an intruder approaches and then suddenly attack it without notice, perhaps gaining some advantage with the element of surprise. The territory is only about one hundred feet in diameter, and nesting always occurs in it, although pair formation seems to take place in the larger area of the range. As the period of egg-laying gets closer, the birds spend more time in the territory, and each one establishes a perch in the area where it frequently stays to rest and preen.

The signal posts (trees where Drumming occurs) are usually spread out over the range. The male and female generally have separate posts, two to four posts for each bird. The Drumming display is easily distinguished from other types of pecking, for it is loud and continuous and no wood is excavated as a result of it. It does not function either to gather food or to excavate a nest hole. This is why some Woodpeckers use gutters or drainpipes as signal posts; it is not because there are insects there or because the birds are making a home, but because the object is so resonant. Drumming seems to serve the function of asserting the bird's presence on its territory. The Khirr-call is also used in this same way.

Courtship

Main behavior: Fluttering-flight, Bill-waving, Still-pose, V-wing, Wing-clapping, Drumming, Tew-tew-call, intrasexual conflicts
Duration: From midwinter until incubation

Courtship among Hairy Woodpeckers takes place in isolated moments of display, occurring occasionally over a very long period. Even though a resident pair will remain together on the same range throughout the year, they interact with each other for only about seven months of the year. This period starts in December or January, and many of the interactions to be seen are described below.

Drumming is one of the first clues that the period of pair formation has started. It may be heard at any time during the day but is most frequent in the early morning. Both male and female drum, and each has its own Drumming trees. If you hear Drumming, locate the trees, for they will help you determine the range of the bird and will often be the sites for further display.

Drumming may be between neighbors or between mates, and in either case one bird may "answer" the other. In the case of mates, the Drumming usually causes one mate to approach the area of the other.

When prospective mates meet, a number of interactions may occur. A common one is the Still-pose, where one or both birds remain motionless for periods ranging from thirty seconds to twenty minutes. These may be preceded or followed by Flutter-ing-flight, a mothlike, hovering flight occurring just before landing or just after takeoff. During the period of pair formation the bird does this display only in the presence of its mate.

Also likely to occur during this period are aggressive encounters with a third bird. The intruding bird will be confronted by the paired bird of the same sex. These conflicts take on certain ritualized forms. Besides Still-poses, Drumming, and Keke-calls, there are the action displays of Bill-waving and V-wing. Bill-waving is the most common and it is done by one or both birds in a conflict. V-wing is a more intense threat and occurs less commonly in these situations. The partner of the defending bird doesn't participate in these interactions, but may follow and feed nearby.

The Bill-wave-call and Bill-waving may also occur to some extent between two birds involved in pair formation. But soon the Bill-waving stops and the Bill-wave-call is replaced by the quieter Tew-tew-call whenever the pair meet. When this latter call is heard between members of a pair, you can be fairly sure that they have formed a bond. From then on, whenever the two come near each other they will give this call. As the pair move about their range, they may also use the Teek-call as a means of keeping contact with each other.

Mating among Hairy Woodpeckers is initiated by the female. She perches at a right angle to a branch (which is a rare position for Woodpeckers) and gives an excited call. The male then steps onto her back and mates. Mating often takes place near the nest site and is most frequent a week or so before egg-laying.

Nest-Building

Placement: 10–40 feet off the ground
Size: Entrance hole 1½–1¾ inches in diameter
Materials: Excavated in live tree

Hairy Woodpecker nest holes are generally excavated in live trees, often aspens in the north and oaks farther south. Both male and female participate in excavation, although the male does the majority of the work. Excavation may be seen at almost any time of the day. The birds often start nest holes in trees and then do not finish them, possibly because the wood is the wrong consistency or because the male has chosen a nest hole in one spot and the female has chosen another. More than other common Woodpeckers, the Hairy Woodpecker seems to choose new nest trees each year. As the birds meet at the nest site, there are often Tew-tew-calls exchanged and the flights to and from the site are often Fluttering-flight.

Locating the Nest

WHERE TO LOOK In wooded areas that have trees with diameters of at least 8 inches

WHEN TO LOOK Begin looking, or listening, for the tapping of excavation from January on.

BEHAVIORAL CLUES TO NEST LOCATION:

1. Look in areas where you see the birds or hear their Drumming.

2. Look for fresh wood chips on the ground and then look up for the nest hole above them.

3. Examine live trees.

Breeding

Eggs: 4–6
Incubation: 11–12 days, by male and female
Nestling phase: 28–30 days
Fledgling phase: A few days
Broods: 1

Egg-Laying and Incubation

During the egg-laying period the adults stay near the nest, and mating is frequent. Eggs are laid in a period of four to six days and during this time the adults establish resting and preening spots on open branches near the nest, each bird taking a separate spot. The adults are very alert around the nest during this phase of nesting.

Both the male and the female take turns incubating the eggs and rarely leave them unguarded. The incubating bird may leave to feed without its mate being near, but this will be for only a few minutes. The male enters the nest hole just before it gets dark and remains by himself on the eggs all through each night. The female then relieves him a little after sunrise, and then they alternate through the rest of the day.

When one adult relieves the other there is often a ceremony. The arriving bird flies to the nest with Swoop-flight or Wing-clapping. It gives the Tew-tew-call (occasionally the Khirr-call) at the nest entrance, waves its head back and forth in front of the nest hole, and then hops to one side. The bird in the nest then comes to the nest hole and leaves with Fluttering-flight.

Nestling Phase

By watching the feeding behavior of the parents, you can estimate the age of the nestlings. When the young are first born, the parent enters all the way into the nest to feed them. Halfway through the nestling phase the young are larger, so the parent can feed them by perching at the nest hole and reaching in with the upper half of its body. In the latest stages the young are big enough to reach the nest entrance, and the parent feeds them without entering the hole at all. After some of the feedings, the parent collects fecal sacs from the nest and carries them away with Fluttering-flight. For the first five days after hatching, the young are brooded by either parent, and during this time the adults may continue with the nest-relief ceremony typical of earlier stages. Once brooding has stopped, the nest-relief ceremony is reduced — the approaching bird giving only the Tew-tew-call and slightly raising its crest feathers. When Starlings are present, the parents guard the nest closely, for Starlings are strong competitors for the nest hole. The nestling phase lasts twenty-eight to thirty days.

Fledgling Phase

The young are well developed by the time they leave the nest and are strong fliers on their first outing. They are fed for a few days outside the nest, begging for food with actions similar to Bill-waving or V-wing. The family stays together within the range of the adults until the young can feed themselves, then the young leave and the parents remain.

Plumage

Hairy Woodpeckers go through one complete molt per year. For adults this may start as early as mid-June and be finished near the end of August. The sexes are easily distinguished by plumage, for the male is the only one to have a red patch on the back of its head. The behavior of both sexes is in most cases very similar.

Seasonal Movement

Most studies show that Hairy Woodpeckers do not migrate, but remain in the same area throughout the year. In midwinter they may drift a few miles away from their range, possibly due to changes in the availability of food.

After the breeding season the first year birds disperse — some moving significant distances to the south, but most moving only a few miles from the place of their birth and remaining there for the rest of their lives.

Feeder Behavior

Hairy Woodpeckers are not as common at feeders as their smaller relative, the Downy Woodpecker. They are especially attracted to suet, but expect only one pair to come to the feeder because of the bird's territorial habits. In early winter the male and female may come separately, while later in winter they are likely to be more closely associated with each other. See the sections on Territory and Courtship.

Most common displays: V-wing and Teak-call. See the Display Guide.

Other behavior: In the vicinity of the feeder you may see courtship or territorial interactions; these would include the displays of Bill-waving, Still-pose, Bill-wave-call, Khirr-call, Drumming, and Keke-call. See the Display Guide and the sections on Territory and Courtship.

Eastern Kingbird / *Tyrannus tyrannus*

THE MOST RENOWNED BEHAVIOR OF THE EASTERN KINGBIRD IS ITS aggressive defense of its territory. Generally, only Crows, Hawks or Owls are persistently attacked. These larger birds can be just crossing through the Kingbird's territory as high as one hundred feet in the air, and still the Kingbird will fly up and continually dive at them from above. Even when the intruder has left the borders of the territory, the Kingbird persists in attacking it. Following these maneuvers in defense of territory, the bird may do a marvelous display called Tumble-flight, where, after flying very high, it glides down in stages, sometimes tumbling in midair. It is interesting that the Kingbird's sense of territory extends so high into the air — I often picture it as a tall cylinder.

There are some curious aspects of the relationship of the sexes in Kingbirds. The male is slightly aggressive toward the female when she first arrives on the territory. This stops after a few days, but then once the female starts building the nest, she does not let the male near it until the eggs hatch. During this period she will even attack him if she returns to the nest and finds him too close. The male typically perches in the nest tree at a proper distance from the nest when the female is off feeding, but he immediately leaves as soon as she returns. Once the eggs hatch, the female allows the male near the nest and at this stage he helps gather food to feed the young birds. Another feature of their relationship is that whenever the two meet during the breeding season, they do the conspicuous displays of Wing-flutter and Kitter-call.

The calls of the Kingbird can be frustrating, for they are hard to distinguish and seem to be extremely variable. A few are listed in the display guide. They are not always heard as distinct calls; instead they are often given in combination. In late summer, when the young have left the nest but are still feeding with the parents, the birds are extremely vocal, so much so that it can be quite bothersome if the family group decides to spend the last month of summer in your backyard.

BEHAVIOR CALENDAR

	TERRITORY	COURTSHIP	NEST-BUILDING	BREEDING	PLUMAGE (MOLTS)	SEASONAL MOVEMENT	SOCIAL BEHAVIOR
JANUARY							
FEBRUARY							
MARCH							
APRIL							
MAY							
JUNE							
JULY							
AUGUST							
SEPTEMBER							
OCTOBER							
NOVEMBER							
DECEMBER							

DISPLAY GUIDE

Visual Displays

Wing-Flutter

Male and/or Female *Sp Su*

The wings are fluttered in shallow arcs.

CALL Kitter-call

CONTEXT May be given while the male is patrolling or whenever two birds within a family group come together after being apart, even for only a short while. *See* Courtship, Breeding

Crouch

Male or Female *Sp Su*

The bird is horizontal and spreads its tail, revealing the white terminal band. The wings are drooped and slightly flicked.

CONTEXT Occurs in territorial defense during the early stages of territory formation; usually directed at an intruder or neighbor

Tumble-Flight

Male *Sp Su*

The bird rises high into the air with stiff fluttering flight. This is interrupted at intervals with sudden tumbles in the air. The flight may end with a fluttering glide to a perch.

CALL Kt'zee-call

CONTEXT Given mostly after attacks on territorial intruders, especially birds of prey; may occur without tumbles, but the fluttering, rising flight is still distinct; also given just before dawn and sometimes before thunderstorms. *See* Territory

Auditory Displays

Kitter-Call

Male or Female *Sp Su*

kitterkitterkitter A rapidly repeated two-part phrase, much like the written description

CONTEXT Always accompanies Wing-flutter, and thus given when two birds of a family group meet after being apart

Kt'zee-Call

Male or Female *Sp Su*

kt'zee. kt'zee. A short, two-part phrase given singly; the first part short, the second part more drawn out

CONTEXT Generally given during aggressive encounters, or by the male when patrolling, or in Tumble-flight. *See* Territory

Zeer-Call

Male or Female *Sp Su*

zeer or *tzit* An extremely short, single sound

CONTEXT Given by male or female in a variety of circumstances, often when the birds are alone; intense versions given in a slow series sometimes precede direct attack. *See* Territory

Dawn-Song

Male *Sp Su*

t't'tzeer, t't'tzeer, t'tzeetzeetzee Two complex phrases continually alternated; complexity and time of occurrence distinguish it from other Kingbird calls

CONTEXT Sung by territorial males during the hour before sunrise; may also be heard in the evening. *See* Courtship

BEHAVIOR DESCRIPTIONS

Territory

Type: Mating, nesting, feeding
Size: Approximately ½ acre
Main behavior: Chases, Zeer-call, Kitter-call, Kt'zee-call, patrolling, Tumble-flight
Duration of defense: From arrival of the male until the start of the fledgling phase

Kingbirds are one of the most aggressively territorial common birds, and yet in many ways their use and defense of territory are not at all like the classic descriptions of territoriality, which are exemplified in the behavior of Song Sparrows or Mockingbirds. When the males first arrive on the breeding ground, they roam over a large area. Then in a few days each begins to restrict his movement to the space of a few acres. He patrols the area in only the morning and afternoon, moving short distances between perches and calling the Zeer-call and Kitter-call. There may be some conflicts between males at this stage if there are many males in a given area.

When the females arrive, usually a few days after the males, the males may at first respond aggressively to them, and since there are very few ways to distinguish the sexes, these may look like territorial skirmishes. Soon a nest site is chosen, and an area with a radius of about thirty yards around the nest becomes the center of their activities.

The territory does not have well-defined borders. Other Kingbirds may enter the territory and not be attacked as long as they do not display toward or try to mate with the female. Birds of other species that are the same size or smaller than the Kingbird are not attacked unless they are very close to the nest. The main reputation for the aggressive behavior of the Kingbird comes from their relations with large birds or birds of prey. Any Crow, Hawk, or Owl that flies near the territory is immediately pursued by either one or both members of the pair. They fly up and repeatedly attack the bird from above while giving the Zeer-call, or Kt'zee-call. They will attack the bird even if it is flying one hundred feet or more overhead, and they keep attacking well past the area of their territory.

After one of these attacks the male may do a marvelous display called Tumble-flight. In this, it flies high in the air with a shallow, fluttering flight, and then descends with a series of short glides sometimes interspersed with acrobatic tumbles. This behavior pattern is variable, but when fully performed it is one of the most exciting bird displays to be seen.

Courtship

Main behavior: Wing-flutter, Kitter-call
Duration: Throughout breeding season

Kingbirds have very few displays or obvious behavior patterns associated with courtship. When the female first arrives, the male may be aggressive toward her, but she will remain in his vicinity, and he soon becomes accustomed to her presence. Whenever the two meet on the territory, there is always a lot of excited displaying involving the Wing-flutter and the Kitter-call. This continues throughout the season and, in its intensity and the amount of energy expended, reminds one of the Greeting Ceremony of Canada Geese.

Other male Kingbirds may enter a territory and attempt to mate with the female, but her mate will be strongly aggressive toward these intruders.

Many observers have noticed a special call, the Dawn-song, given by the male Kingbird. It is characterized as the alternating of two fairly complex phrases and may be heard on any day of the breeding season, especially during the hours just before dawn and to a lesser extent at dusk.

Nest-Building

Placement: Most often near the tip of a horizontal branch, 10–20 feet high; sometimes on top of stumps in water or on fence posts in open country
Size: Inside diameter, 3 inches
Materials: Loosely made of soft materials such as weeds, moss, bark strips, feathers, cloth and string

Kingbirds usually return to the same nesting area throughout their lives. The female constructs the nest by herself. Although the male may accompany her as she gathers material, he generally perches near the nest while she is away. When the female returns, the pair greet each other with Wing-flutter and Kitter-call. The female is extremely protective of the nest and the immediate area around it, so much so that she doesn't even let the male perch on the nest. This relationship continues until the nestling phase, when both male and female approach the nest to feed the young.

The nest is a very messy conglomeration of soft, fibrous materials, often with long strands hanging off the edges, and it is easily destroyed by a good thunder shower. The birds may build several preliminary nests before building the final one for breeding.

Locating the Nest

WHERE TO LOOK Near open water or in orchards or overgrown fields

WHEN TO LOOK From the arrival of the female in late spring until the completion of the nestling phase in midsummer

BEHAVIORAL CLUES TO NEST LOCATION:

1. Listen for any Kingbird calls, for the nest is most likely nearby.
2. Watch for the birds hunting aerial insects from an exposed perch, for these are usually located near the nest site.
3. Watch for food being brought back to the nest, for the birds are not secretive about approaching the nest.

Breeding

Eggs: 3–4
Incubation: 14–16 days, by female only
Nestling phase: 14–17 days
Fledgling phase: Variable, often 2–3 weeks
Broods: 1

Egg-Laying and Incubation

After the completion of the nest there is a period of three to four days before egg-laying begins. The eggs are laid one per day until the clutch is complete. Only the female incubates, and she leaves the nest about once every half hour to feed for five to ten minutes. The male is kept away by the female from the immediate area of the nest throughout this phase. He feeds elsewhere while she is incubating, and then, when she leaves to feed, he frequently perches in the nest tree until she returns. When she returns the two do Wing-flutter and the Kitter-call as they meet, and then the male goes to another part of the territory to feed.

Nestling Phase

Once the eggs have hatched, the relationship between the male and the female begins to change. For the first few days she broods the young and is still aggressive if the male approaches the nest. But after that she allows him near the nest, and both parents begin to share in the feeding of the nestlings. Fecal sacs are carried out by both parents and often dropped from certain perches, resulting in quite an accumulation underneath these spots.

The duration of the nestling phase is variable, for the young may at first move to branches near the nest before they can fly. The phase may last from fourteen to seventeen days.

Fledgling Phase

This phase may last a couple of weeks and merges gradually

with the time when the young are independent. The family stays together for a long time, often remaining in one area, and is believed to break up just before migration.

While the young are still fledglings, the birds are very conspicuous. Both the young and the adults give loud Kitter-calls whenever food is exchanged, and the adults still do Wing-flutter and voice Kitter-calls when they meet.

Plumage

The molts of the Eastern Kingbird are not well understood. The adult birds do not seem to molt while on the breeding ground. It is believed that after their fall migration to South America, they go through one or two molts before returning the next spring.

It is difficult to distinguish male and female Kingbirds by either appearance or behavior. The most obvious clues are that the female is the only one to build the nest and incubate the eggs.

Seasonal Movement

Eastern Kingbirds migrate to South America for the fall and winter. The fall migration occurs in late August and the birds return north during March and April. The birds generally migrate in widely dispersed small flocks.

Tree Swallow / *Iridoprocne bicolor*

THE BEAUTIFUL FLIGHT OF TREE SWALLOWS AS THEY SWOOP IN graceful arcs to gather food is enough to fill many delightful hours of observing. But if you watch the birds over a period of a few weeks, you are bound to see some unusual occurrences that will make you wonder more deeply about their lives. Imagine that you have visited a Tree Swallow nesting area for a few days and found the birds actively feeding, nest-building, and courting. You may return the next day at the same time to discover that there is not a single swallow to be seen. They may stay away for up to four days, and then, just as suddenly, return. No one has been able to account for these unexpected departures or explain where the birds go.

Other aspects of Tree Swallow behavior are the interesting displays around the nest site. Two common visual displays are Flutter-flight, done by the male in the presence of the female, and Bowing, which is done together by the pair very near the nest site, or in the case of nesting boxes, right on top of the box. A third display, which is sometimes harder to see, is Billing, when the male and female touch beaks.

An especially good place to observe Tree Swallows is in conservation areas where there are many nesting boxes, for here the Swallows will be active and you will have more chances to see interactions. The birds are very active in defending their nests, and at certain times in the breeding season, if you approach too near, a whole group of birds from the area will give the Cheedeep-call and start circling above you. Some of the birds, possibly the owners of nearby nests, will actually do spectacular dives at your head, swerving just at the last moment and giving a call as they pass. A similar pattern of behavior, called Towering (*see* Social Behavior), occurs early in the season, but at these times it is not in response to predators and involves the whole nesting colony. Its function is still to be discovered.

BEHAVIOR CALENDAR

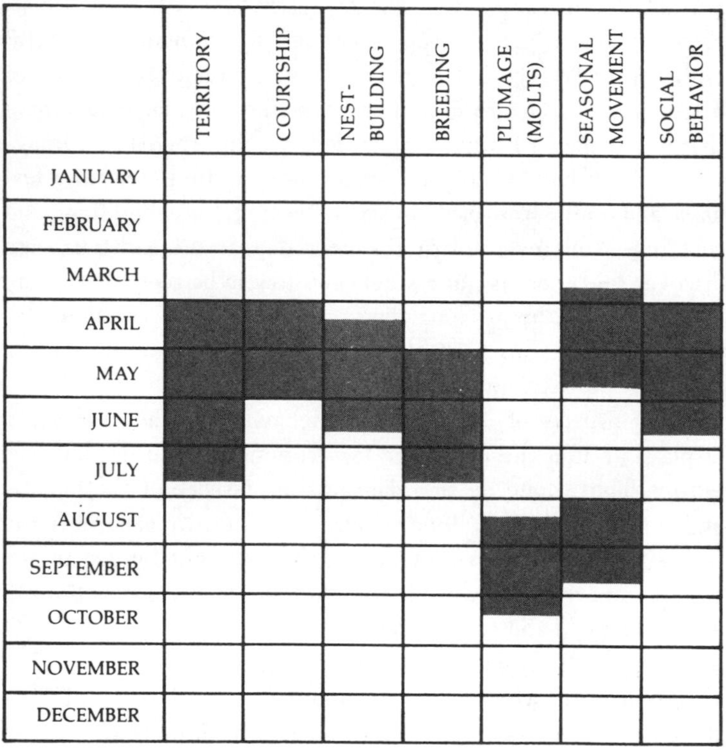

	TERRITORY	COURTSHIP	NEST-BUILDING	BREEDING	PLUMAGE (MOLTS)	SEASONAL MOVEMENT	SOCIAL BEHAVIOR
JANUARY							
FEBRUARY							
MARCH							
APRIL							
MAY							
JUNE							
JULY							
AUGUST							
SEPTEMBER							
OCTOBER							
NOVEMBER							
DECEMBER							

DISPLAY GUIDE

Visual Displays

Flutter-Flight

Male *Sp Su*

The male hovers or flies short distances with rapid, shallow wingbeats. He sometimes approaches the female with this flight.

CALL Twitter-call

CONTEXT Usually done near female; often followed by the bird's landing near or on the nest and doing the Bowing display; may also be followed by mating. *See* Courtship

Bowing

Male and Female *Sp Su*

Two perched birds approach each other, rhythmically bowing the front halves of their bodies toward the ground.

CALL None or Song

CONTEXT Done by a pair near their nest. *See* Courtship

Billing

Male and Female *Sp Su*

Two birds touch bills.

CONTEXT Occurs between mates near the nest. *See* Courtship

Auditory Displays

Song

Male *Sp*

Three fairly long, descending notes, followed by a warbling call

CONTEXT Done by the male, especially when he is near his mate; may accompany Bowing and is heard from his arrival on the breeding ground until the incubation phase. *See* Courtship

Cheedeep-Call

Male or Female *Sp Su*

cheedeedeedeedeep A sharp, rapidly repeated note

CONTEXT Given when there is possible danger near the nest. The birds generally circle over the intruder's head while giving the call. A somewhat softer version of this call may cause "Towering" early in the breeding season. *See* Breeding, Social Behavior

Twitter-Call

Male *Sp*

A twittering call given during Flutter-flight

CONTEXT Usually precedes mating. *See* Courtship

BEHAVIOR DESCRIPTIONS

Territory

Type: Nesting
Size: Immediate area of the nest
Main behavior: Chases

Duration of defense: From arrival on the breeding ground until the fledgling stage

Tree Swallows defend only the nest itself. They are tree-hole-nesting birds, which have readily taken to using man-made nesting boxes. There is a lot of competition for this type of nest, and Tree Swallows can be very aggressive even against their own species when trying to claim a nest. Other species that compete with them are House Sparrows, Bluebirds, and House Wrens.

Both the male and the female help defend the nest, and when you or any other possible predator approaches it, the birds will give the Cheedeep-call and circle over your head. Other birds in the immediate area will join in, and every once in a while one will dive directly at your head. This can be unnerving, for they swerve off only at the very last moment. When competing with its own species the defender may meet with the challenger in midair, and the two grapple as they begin to fall.

Courtship

Main behavior: Flutter-flight, Bowing, Billing, Song, Twitter-call
Duration: 1–2 weeks

Soon after the Tree Swallows arrive on the breeding grounds you will see pairs of birds perched near their nest holes or on top of their nest boxes. You will hear the male give Song frequently until the beginning of the incubation period. Sometimes the male will do Flutter-flight about the nest box when the female is there, and then land and move toward her doing the Bowing display as he gets closer. Billing may also be seen in these situations, the male and the female touching beaks.

Mating starts to occur about a week before egg-laying begins. Before mounting the female, the male goes into Flutter-flight about her. Then he begins to descend while giving the Twitter-call. If the female is receptive she leans forward, and the male lands on her back, holding her head feathers in his beak and bending his tail down to make contact with her. This whole pattern may then be immediately repeated. If the female is unreceptive to the male, then she will raise her wings in such a way that he does not land. Most courtship displays are stopped once incubation has started.

Nest-Building

Placement: In a tree hole or nest box about 5–10 feet above ground
Size: Entrance hole 1½ inches in diameter
Materials: Grasses and then a lining of feathers

Nest-building for Tree Swallows can be a very slow process; the birds often take up to a month to complete their nests. The average construction time is two to three weeks.

The female does the majority of building although the male may help gather materials. The nest is made of grasses built into a foundation and then is lined with feathers. Sometimes there is competition for these feathers, and you may see some Swallows trying to steal feathers that others have collected.

Locating the Nest

WHERE TO LOOK In open areas where there are nest boxes or old tree holes; often near water or in dead trees at the water's edge
WHEN TO LOOK From when the birds first arrive until midsummer
BEHAVIORAL CLUES TO NEST LOCATION:

1. The Swallows often feed over a nesting area.
2. When an intruder approaches the nest, the birds circle close overhead and give the Cheedeep-call.

Breeding

Eggs: 5
Incubation: 14–15 days, by female only
Nestling phase: Variable, but averages 21 days
Fledgling phase: 2–3 days
Broods: 1

Egg-Laying and Incubation

The eggs are laid one each day until the clutch is complete, except in cases where there is a temporary abandonment of the nesting area (*see* Social Behavior). Egg-laying generally starts right after the nest is completed, but with first-year females, it may start earlier, and if it does, the birds will continue to line the nest during the egg-laying period.

Incubation is done only by the female, but the male guards the entrance to the nest when the female leaves to feed. When the female is about to leave the nest, she waits with her head and shoulders outside the nest hole. When the male arrives, she leaves and he enters the nest and immediately turns around to place his head and shoulders in the nest hole and peer out. He leaves when the female returns, and she continues incubation. Occasionally the male brings food to the female while she is incubating.

During incubation the adults may temporarily abandon the nest area for several days, leaving the eggs uncovered and unprotected. This behavior makes the incubation period longer.

Nestling Phase

The young are brooded by the female for the first three days of the nestling phase. Both parents participate in feeding and make frequent trips to the nest. They may go up to a mile from the nest to gather food. The adults have the interesting habit of carrying the fecal sacs of the young away from the nest and dropping them into water. They will do this even if the water is a few hundred yards away, but if it is any farther than that they will just drop the sacs on the ground.

Fledgling Phase

There is practically no fledgling phase. The young are strong fliers as soon as they leave the nest, and having left it, they do not return. They may, however, be fed by the parents in the vicinity of the nest for two to three days before being totally independent.

Plumage

Adult Tree Swallows go through one complete molt per year. This occurs in late summer just before migration. In this respect they differ from most other Swallows, which generally go through their molt after migration.

Young males in their first year already have the iridescent blue-green back of older adults, but females may not acquire this coloring until their second or third year. Before that they have a slightly browner back. Because of this, females in their first two years can be distinguished from males and from older females, but after that the sexes are identical in appearance.

The best behavioral clues for distinguishing the sexes are that the male is the only one to give the Song display and the female is the only one to incubate the eggs.

Seasonal Movement

After the fledgling phase, the young and adults may leave the breeding area and join other Swallows over marshy areas where aerial insects are plentiful. They also roost in these areas, and, before settling in for the night, may funnel down in large formations into the roosting trees.

In early fall, Tree Swallows start migrating to the south. There they remain as large flocks through the winter, feeding and roosting in marshy areas. In spring they migrate north in much smaller flocks that arrive on the breeding grounds over a period of a month.

Social Behavior

A fascinating aspect of the life of Tree Swallows is their habit of occasionally abandoning their breeding ground. It can be a real shock if one day you visit a nesting area and see tens of Swallows active about the nest boxes, and then on the next day you find no Swallows at all. It is not clear what the Swallows do when they leave the breeding area, but it generally occurs on cool, cloudy days when there is a lack of aerial insects, on which the birds depend for food. They may interrupt egg-laying for up to four days during one of these abandonments and then return to continue laying the clutch. Even more remarkable is the fact that this has little effect on the overall success of the eggs; these abandonments may even occur in the middle of the incubation phase with no ill effects on the young. During these periods observers have gone out to look elsewhere for the Swallows, but have not been able to find them.

Another interesting group activity of Tree Swallows is called "Towering." This behavior occurs mostly in the period before egg-laying has started. It starts with one bird's flying far above and giving a call similar to the Cheedeep-call. The other Tree Swallows in the area join the first, and circling high in the air they give the call for a minute or two and then drop down to their nest areas. This is slightly different from the response to predators, which generally involves a few birds who fly directly over the danger and dive down at it.

Both the function of Towering and the reasons for mass abandonment of the breeding area are still intriguing mysteries of the Tree Swallows' behavior.

Blue Jay / *Cyanocitta cristata*

WITH JUST CASUAL OBSERVATION OF BLUE JAYS, YOU CAN DISCOVER some striking changes in their general behavior. During fall the birds gather into large flocks that roam about and feed. By midwinter these flocks usually divide up into smaller groups that remain only loosely associated with each other. In spring Blue Jays enter their loudest and most active stage, flying about in the mornings in a follow-the-leader fashion and giving their greatest variety of calls. Following this there is about a month of almost total quiet when mated pairs are involved with mate-feeding, nest-building, and incubation. Once the young have left the nest, the family group is often seen traveling about together, and the young often call to elicit food from the parents.

The two most contrasting stages are in spring and early summer. The early courtship flocks of spring are very conspicuous. They are believed to be composed of one female and a number of males, the female always in the lead as the birds take flight. After the flock lands, the males display, one of the more common displays being Bobbing. (*See* display guide.)

After this stage is a fascinating period that could be easily overlooked. Mated pairs become extremely quiet, remain near the areas of their future nests, and skulk through trees and underbrush like ghosts of their former selves. One way to locate a pair is to listen for the Kueu-call. This is given by the female as the male approaches her with food. Mate-feeding starts during this quiet phase of the birds' lives and continues on through the incubation period.

The calls of the Blue Jay are extremely varied, and it is often hard to determine which calls are distinct and which are just variations of the same call. Only a few of the commonly repeated calls are listed in the display guide; more would have been included were it not for the surprising fact that very little research has been done on this marvelously expressive bird.

BEHAVIOR CALENDAR

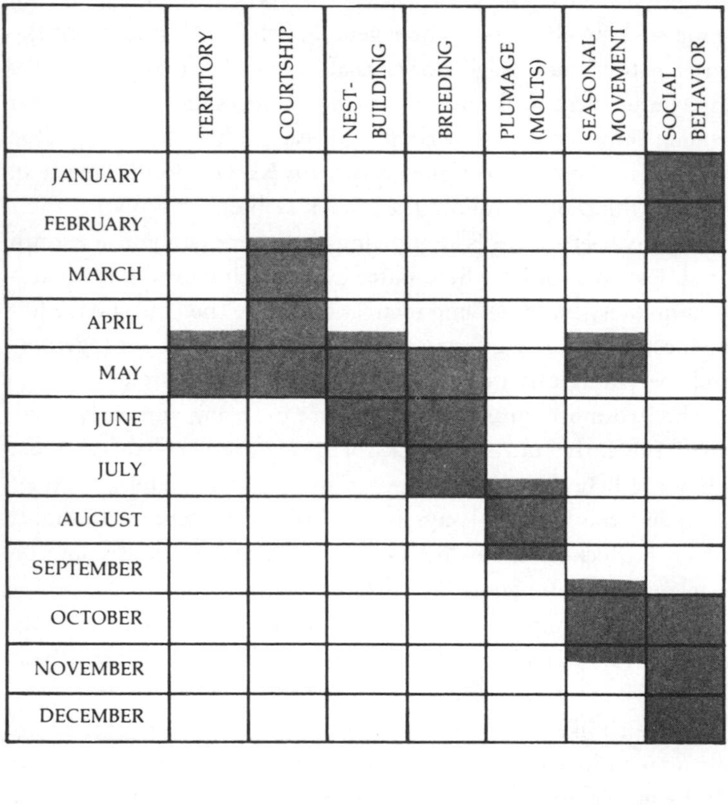

	TERRITORY	COURTSHIP	NEST-BUILDING	BREEDING	PLUMAGE (MOLTS)	SEASONAL MOVEMENT	SOCIAL BEHAVIOR
JANUARY							
FEBRUARY							
MARCH							
APRIL							
MAY							
JUNE							
JULY							
AUGUST							
SEPTEMBER							
OCTOBER							
NOVEMBER							
DECEMBER							

DISPLAY GUIDE

Visual Displays

Bobbing

Male or Female *Sp Su F W*

The bird quickly raises and lowers its whole body repeatedly by extending its legs.

CALL Toolool-call

CONTEXT Occurs during courtship flocking and during aggressive encounters; often done in courtship flocks by more than one bird at a time. *See* Territory, Courtship

Body-Fluff

Male or Female *Sp Su*

A bird crouches down, fluffs its body feathers, and holds its crest erect. The head may be slightly lifted.

CONTEXT Done by birds that are intruding on another's territory and are being confronted by the territory holder. *See* Territory

Auditory Displays

Jaay-Call

Male or Female *Sp Su F W*

A raucous call that has many variations and that can be given at varying intensities

jaay. jaay.

CONTEXT At lower intensities, used as an assembly call that attracts other Jays, as in courtship flocking; at higher intensities, used as an alarm or mobbing call. *See* Courtship, Breeding, Social Behavior

Toolool-Call

Male or Female — *Sp Su F W*

toolool, toolool,

A bell-like call of two parts, both on the same pitch

CONTEXT Occurs with Bobbing; directed at other males in the courtship flocks and at predators. *See* Courtship

Rattle-Call

Male or Female — *Sp Su F W*

A dry, nonmusical rattle

CONTEXT Occurs at times with Bobbing or alone, and in a variety of situations; particularly evident at time of courtship flocks. *See* Courtship

Kueu-Call

Male or Female — *Sp Su F*

kueu kueu kueu or *kuetkuetkuetkuet*

A softly repeated sound fairly similar to its written description

CONTEXT Given during courtship when male feeds female, or during nest-building. *See* Courtship, Nest-building

Wheedelee-Call

Male or Female — *Sp Su F*

wheedelee

Has been aptly described as resembling the sound of a squeaky gate

CONTEXT Given during early courtship flocking. *See* Courtship

BEHAVIOR DESCRIPTIONS

Territory

Type: Nesting
Size: Not well defined
Main behavior: Aggressive interaction with intruders
Duration of defense: From nest-building to fledgling phase

There is very little noticeable territorial behavior in Blue Jays. From nest-building to the fledgling phase there appears to be some defense of an area against other Jays, but even at this time, defense seems to be directed only at Jays that are in the same breeding stage. Other Jays are freely permitted to land in the area of the nest even when young or eggs are there.

Interactions between a territorial bird and an intruder take on a certain form. The intruder or subordinate bird assumes the Body-fluff display, crouched low with crest raised and bill held slightly upward. The resident bird approaches the intruder with crest down and head forward. It hops about the other bird, and a short chase or skirmish may follow. The Bobbing display may also be used in these situations by the dominant bird.

Blue Jays live within a given range through most of the year. During the breeding season the ranges of neighboring birds may broadly overlap, and during fall and winter certain groups of Jays seem to gather together in some ranges.

Although Blue Jays show little inclination to defend their nests against other Jays, they can be extremely persistent in defending them against possible predators. They react particularly strongly to cats and squirrels, and a pair will continually dive at these enemies and give loud Jaay-calls until the danger has moved twenty or thirty yards away.

Courtship

Main behavior: Mate-feeding, Bobbing, Kueu-call, Toolool-call, noisy flocks
Duration: February to June

Pair formation in Blue Jays can be divided roughly into two phases — an early phase, which consists of actively displaying flocks, and a later phase, which is quieter and consists of pairs or groups of three birds.

The early courtship flocks are a conspicuous aspect of Blue Jay behavior in late winter or early spring. They consist of three to ten birds actively flying about as a group. This group forms in the early morning, roams an area of about one-quarter of a square mile, and usually disbands by midday. The flocks are believed to be composed of one female and the rest males. The female takes the lead in all of the flock's activities. If she flies, the rest follow; if she lands, the rest land; if she is still, the others are still; if she hops higher in a tree, so will the rest. Certain displays typically occur within these groups, and these actions help one to distinguish a courtship flock from just a random assortment of Jays earlier in the season.

The flock is usually raucous in flight and lands together in the tops of trees. Once they have landed, the birds may become still for a moment. Then if one starts the Bobbing display, the others may join in for a few seconds of vigorous Bobbing. This is often accompanied by the Toolool-call or the Rattle-call. The Wheedelee-call is also given while the birds are perched; and as they fly off to a new perch, numerous variations of the Jaay-call are heard.

As the season progresses these flocks grow smaller until there are only two or three birds in a group. What is believed to be happening is that the males are in some way competing among themselves for the female, and that many of the displays associated with the flocks are aggressive or intimidating enough to make some birds drop out of the chase.

The second phase of courtship is a striking contrast to the earlier period of noise and conspicuous activity. Birds, once paired, move about in stealth and silence, and stay independent of younger groups of Jays that are still noisy. During these quiet times together, the female can be heard giving the Kueu-call; she usually does this when the male is feeding her. The female may fluff out her feathers as she receives the food, or she may receive it with little display other than the call. The transfer of food is brief, with the male hopping toward the female, touching his bill to hers, and then hopping off. The ceremony occurs frequently throughout the day and is a common sight during this phase of courtship.

It is believed that the early courtship flocks are composed primarily of birds pairing for the first time. Birds that have bred in previous years join into pairs earlier in the season and also start their mate-feeding at that time.

In May and June you may see a renewal of courtship activity. The birds in these flocks are mostly first-year birds that do not breed during the coming season.

Mating occurs during the last days of nest-building.

Nest-Building

Placement: Usually in an evergreen, from 10 to 20 feet above ground
Size: Inside diameter 3½–4 inches
Materials: Twigs, bark, leaves and many man-made objects such as plastic, string, cloth, paper; lined generally with fine rootlets

Blue Jays frequently build preliminary nests that are not completed or used for breeding. These are sometimes built in an entirely different manner from the breeding nest. The twigs are gathered by the male only and are not taken from the ground but broken directly off trees. As the male offers the twigs to the female at the nest site, she gives the Kueu-call and then arranges them into the nest. The preliminary nest never goes beyond the stage of being a loose platform of twigs, and several may be built before the breeding nest is started.

Both male and female collect twigs from the ground for the breeding nest, and both participate in its construction, although the female does most of the work. About five days are needed to complete the breeding nest.

Locating the Nest

WHERE TO LOOK In areas of mixed scrub and mature growth, and in either urban or natural locations

WHEN TO LOOK Mid-spring, when second stage of courtship is under way

BEHAVIORAL CLUES TO NEST LOCATION:

1. The only real clue is to follow the activities of quiet pairs early in the morning.
2. Follow birds carrying nesting material.
3. Remember that the nest being built may be only a preliminary nest and might be abandoned in a day or two.

Breeding

Eggs: 4–5
Incubation: 17 days, by female only
Nestling phase: 17–19 days
Fledgling phase: Up to two months
Broods: 1–2

Egg-Laying and Incubation

During the last stages of nest-building, the female may sit on the nest for long periods even before egg-laying has begun. Once the eggs have been laid and real incubation has started, she receives all of her food from the male. At first she leaves the nest briefly and is fed nearby, but in the later stages of incubation she takes all her food right at the nest. Her uninterrupted periods of incubation get longer as the incubation phase progresses, until, near the end of it, she may leave the nest for as little as a few minutes every two hours. During this phase the male generally stays near the nest and guards against any possible danger.

Nestling Phase
The female broods the young for the first few days after they hatch. Both parents take part in feeding them, and the male also continues to bring food to the female. The young remain in the nest for a comparatively long period and are well grown by the time they leave. During the last days of this phase, the young may stand on the edge of the nest and exercise their wings.

At the start of the nestling phase the parents become more vocal and conspicuous as they defend the nest against any possible predators. They will dive at cats especially and give variations of the Jaay-call until the cats are far away from the nest.

Fledgling Phase
The family group stays together for a long time after the young have fledged, and the young, even when as large as the adults and seemingly able to gather food for themselves, still receive food from the parents. Their begging calls become extremely raucous and can be heard even into September. These noisy bands of young birds are in striking contrast to the older Jays, which are quiet during the later stages of courtship and the incubation phase.

Plumage

Blue Jays go through one complete molt per year. This starts at the end of July and continues through August. There is little change in the birds' appearance through the year except that after the molt their feathers are brighter in color and reveal more black barring. Just before the molt the birds' feathers are worn at their tips and the brightness and distinct barring are gone.

The male and female are identical in appearance, and their

behavior is also quite similar. Only in spring and summer will you have some clues to the sexes, for before the young have hatched, the female is usually the one to receive food during mate-feeding. Also, the female is the only one to incubate the eggs.

Seasonal Movement

Most Blue Jays stay within the same areas throughout the year, drifting slightly to new locations in fall and winter if their demands for food cannot be met. Other Blue Jays, however, migrate south in fall and north in spring. The former tend to be the adults, while the latter are primarily first-year birds.

Social Behavior

Parents and young generally stay together through summer and in fall. They sometimes join other family groups and lone birds in areas where there is ample food. These larger flocks of Jays are typical in fall, and they often spend a great deal of time in the tops of oaks feeding on the acorns. When subfreezing weather sets in, these flocks seem to disperse into smaller groups averaging four to six birds, which then become the normal grouping for most of winter.

In late winter large groups of Jays are again seen, sometimes feeding together in the early morning and then dispersing. These may be precursors to the early courtship flocks.

Blue Jays, like their close relative the Common Crow, have a tendency to mob birds of prey such as Hawks or Owls. The first bird that sees the predator gives the Jaay-call, and this attracts other Blue Jays in the area. The Jays remain perched around the predator, calling loudly and diving down upon it but rarely actually touching it. This may continue for half an hour or more, and generally ends with the predator's moving away or the Jays' gradually ending their mobbing and dispersing.

Feeder Behavior

Blue Jays are often regular visitors to feeders, especially those that offer food on the ground. They seem to prefer the larger seeds, such as sunflower and cracked corn. They may take many seeds all at once or just take one, fly off with it, and return for another. In fall Jays wander about in large flocks; in winter they form smaller groups that tend to remain in a given area. See the section on Social Behavior.

Most common displays: The Jaay-call is often given. It usually startles other birds at the feeder. See the Display Guide.

Other behavior: In spring you may see some of the courtship behavior that takes place in flocks. Displays in these situations include Bobbing, Toolool-call, Rattle-call, and Weedelee-call. A little later you may see the male feed the female at the feeder. This is a later stage of courtship. See the section on Courtship.

Common Crow / *Corvus brachyrhynchos*

FROM FALL THROUGH WINTER, CROWS ARE USUALLY IN LARGE, raucous flocks that roam widely, but from spring through summer they are more often in small bands, spending the majority of their time in fairly restricted areas. These two broad patterns of behavior reflect the nature of the Crow's nonbreeding and breeding periods respectively.

During the nonbreeding period the most obvious feature of Crow behavior is their habit of gathering into huge communal roosts. The birds are believed to return to the same roost each night, and their behavior as they approach the roost is often predictable. They may fly to the roost along certain established flight lines. If you are near one of these, you will notice that for as long as an hour in the late afternoon Crows will be passing overhead, a few at a time. These flight lines may end at preroosting spots, where the birds first gather before heading to the primary roost. The Crows may be very active at the preroosting spot, calling loudly and diving among the trees as they chase after each other.

In spring Crows behave differently. Small groups, believed to be composed of a breeding pair and their nonbreeding young from the previous year, fly about together but remain in a fairly restricted area. At first these groups are noisy and fight with other Crows, but soon after that there comes an extremely quiet time when the birds become secretive. This lasts for about a month and is a striking occurrence for anyone who has been following the birds closely. It happens during the nest-building, egg-laying, and incubation periods. Soon after this the parents become noisy defenders of the nest, and you will hear the wailing call of the young as they elicit food from the parents.

There has been amazingly little study of Crows by researchers, so most of the best questions about the birds still cannot be answered. What is included in this guide is much of what is presently known, and it is hoped that you can use this as a basis for your own exploration of these birds' intriguing behavior.

BEHAVIOR CALENDAR

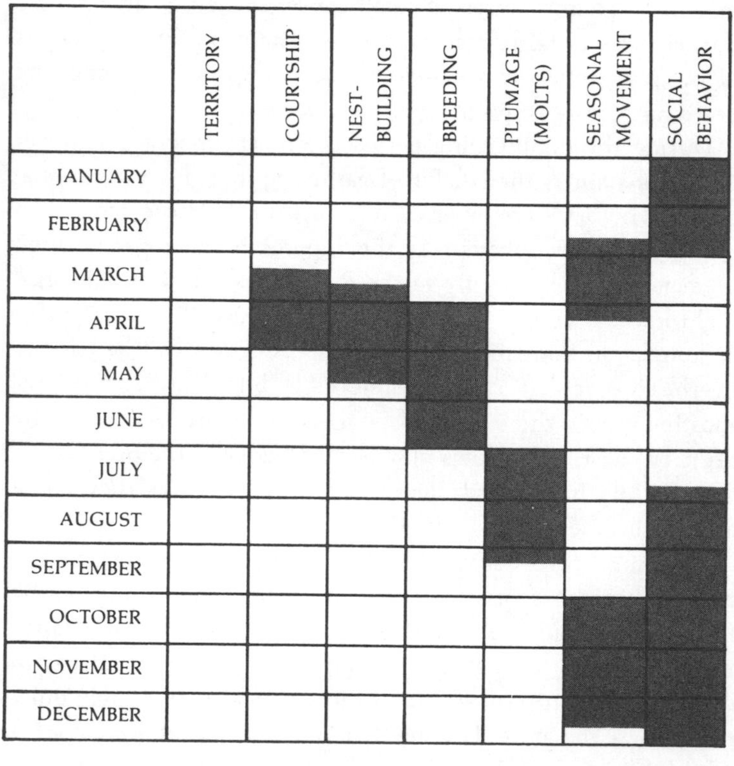

	TERRITORY	COURTSHIP	NEST-BUILDING	BREEDING	PLUMAGE (MOLTS)	SEASONAL MOVEMENT	SOCIAL BEHAVIOR
JANUARY							
FEBRUARY							
MARCH							
APRIL							
MAY							
JUNE							
JULY							
AUGUST							
SEPTEMBER							
OCTOBER							
NOVEMBER							
DECEMBER							

DISPLAY GUIDE

Visual Displays

Bobbing

Male or Female *Sp*

The bird bobs its head up and down, and may accentuate this by bowing the whole front of its body. The wings and tail may be spread slightly and the body feathers may be fluffed.

CALL Rattle-call

CONTEXT Usually given in the presence of another Crow; most often seen in spring and thus possibly associated with courtship. *See* Courtship

Auditory Displays

Rattle-Call

Male *Sp*

A dry, rattling call, very different from the normal crow *caw*

CONTEXT Usually occurs with the Bobbing display and is heard almost exclusively in spring. *See* Courtship

The great variety of other sounds that Crows make cannot be sufficiently distinguished in this book to enable the reader to identify them in the field. It is still questionable how many distinct calls there are and how much of the difference among calls is due to individual variation.

BEHAVIOR DESCRIPTIONS

Territory

Type: Nesting
Size: Immediate area around nest
Main behavior: Chases
Duration of defense: From the start of nest-building to the fledgling phase

Common Crows are not believed to be territorial, but they may occasionally keep certain other Common Crows away from their nesting trees. Crows participating in the nest defense may include extra birds over and above just the parents. For an explanation of this, *see* Breeding.

Courtship

Main behavior: Bobbing, Rattle-call, fighting among members of small flocks
Duration: A few weeks in spring

Courtship is not a prominent aspect of Crow behavior. In late winter you will notice that the large flocks of Crows typical of fall and early winter no longer appear, but instead, Crows are seen in small groups of four to five birds, and sometimes just in pairs. Chases and fairly serious fights among the members of these groups are common, often with two birds grappling as they fall through the air. The only behavior resembling courtship displays is Bobbing accompanied by the Rattle-call. In this display one bird faces another and bows deeply while spreading wings and tail. The body feathers may also be fluffed and the Rattle-call is sometimes followed by soft *coo*ing notes. The Rattle-call is given primarily in the spring and is not heard at most other times of the year. Because of this timing, many observers believe that it is associated with courtship. These displays occur most often in the very early morning.

Nest-Building

Placement: Generally high in trees, 20–60 feet above the ground
Size: Outside diameter 24 inches; inside diameter 7 inches
Materials: Sticks and twigs, bark strands; lined with soft materials such as shredded bark, grasses, wool, moss, string or cloth

A good sign that nesting is about to begin is a small group of Crows that remains in a particular area day after day. If this occurs, begin to look for birds carrying nesting material in their beaks, for this can help you to locate the nest. Both the male and the female participate in collecting material and building the nest. Many of the larger twigs that form the base are broken directly off trees, so if you see a Crow hopping slowly about some dead branches, continue to watch and you may see it break off a branch and carry it to the nest.

Crows may return to their same nesting tree two or three years in a row, and they often return a few weeks before they start building. Crows are common birds, but their nests are not as easy to find as one would expect. This is made even more difficult by the fact that they may partially construct a number of preliminary nests. Nest-building may occur to a limited extent for a month or more, the final nest being completed in one to two weeks.

Locating the Nest

WHERE TO LOOK In tall trees near clearings, often in the tops of pines or oaks

WHEN TO LOOK In late winter and early spring for nest-building; later, through observation of breeding activities

BEHAVIORAL CLUES TO NEST LOCATION:

1. Look for small groups of Crows frequenting an area where there are possible nesting trees.
2. Watch for Crows carrying large twigs or other nesting material.
3. Look for strangely quiet and secretive behavior in the area of possible nesting trees.

Breeding

Eggs: 4–5
Incubation: 18 days
Nestling phase: 4–5 weeks
Fledgling phase: 2 weeks
Broods: 1–2

Egg-Laying and Incubation

An unusual aspect of Crow breeding behavior is that there are most often three to five birds associated with a particular breeding site. It is believed that the extra birds around an adult pair are their young from the previous year. These extra birds, which do not breed until their second year, help to defend the nest area, but it is not yet known in what other ways they participate in the breeding cycle. Some observations suggest that they may help feed the female while she incubates the eggs, or even help feed the young when they are hatched.

The eggs are laid soon after the nest is completed. One egg is laid each day until the clutch is complete. Incubation is done by only the female; the male remains in the area of the nest and occasionally brings food to the female at the nest. More often she leaves the nest to feed. The egg-laying and incubation period is marked by extremely quiet and secretive behavior of the breeding adults, and this makes it very hard to locate their nests at this stage. As with Blue Jays, the birds become conspicuous to the behavior-watcher precisely because their habits are so changed.

Nestling Phase

The nestling period is four to five weeks long, and the young are brooded by the female for the first ten days. Both parents feed the nestlings, but whether the first-year birds help with the feeding is still not clear. The first-year birds are allowed near the nest and may even perch on its rim. The parents become increasingly vocal as the nestling phase progresses, and the nestlings also get louder when they beg for food during the later stages. In the last week of the nestling phase, the young may perch at the edge of the nest or even on nearby branches as they are fed by the parents.

Fledgling Phase

For the first few days after fledging, the young remain in the general area of the nest. Once they can fly well, they begin to follow the parents about and beg loudly for food. The typical sound the young make when receiving food starts off like a begging call and ends up sounding as if the birds were gargling or choking. Hearing this call is a good way to determine what phase of their life cycle the Crows are in. The fledgling phase lasts only two weeks, for after this the young feed on their own, even though they may stay with the parents throughout the day.

Plumage

Adult Crows go through one complete molt per year, and this occurs in summer. There is no difference in the appearance of the male and the female, and the only behavioral clue is that the female is believed to do most of the incubation.

Seasonal Movement

The Crow's seasonal movement patterns differ in various regions of the continent. Long distance migration is only known to take place in the Midwest, the birds moving from the Canadian provinces one thousand to two thousand miles down into the midwestern states. In other northern areas Crows seem to move at most only a few hundred miles in spring and fall. In the southern part of the country, the Common Crow is a year-round resident.

For all Crows there is a general drifting after the breeding season to areas where there is abundant food, and in these places the birds gather into large flocks, which in turn gather with other flocks at night in immense communal roosts. *See* Social Behavior.

Social Behavior

Outside of the few months of the breeding season, Crows are extremely gregarious. After the last young have fledged, the family group usually joins other groups of Crows, and these begin to form a large flock that divides up for feeding during the day, but gathers again each night to roost. The roost becomes an important focal point in the birds' life outside the breeding season. Each morning the roost breaks up into smaller flocks that disperse across the countryside to feed. Some flocks may fly up to fifty miles from the roost each day. In midafternoon these smaller flocks start back toward the communal roost. They fly along fixed flight lines used each day and are joined by other flocks as they go. Often there are preroosting sites, where flight lines coincide and Crows stop to feed before making the final trip to the roost. At these spots there may be much chasing and often spectacular dives as the returning Crows join the others at the preroosting spot. Then just before dusk all the Crows in the area enter the roost site together. These last flights into the roost can be spectacular, for they may contain anywhere from a few hundred to a few hundred thousand birds. The largest roosts are where birds migrating from the north come together with local, year-round residents.

Another social behavior of Crows that occurs throughout the year is their habit of mobbing birds of prey, especially Owls. The long, excited calls of the Crows as they mob an Owl are very characteristic and are good to follow up, for they usually lead you to the sighting of one of our larger Owls.

Black-Capped Chickadee / *Parus atricapillus*

THE BLACK-CAPPED CHICKADEE IS ONE OF THE BEST BIRDS FOR learning about the various uses of auditory displays. You can often tell what a Chickadee is doing just by hearing one of its calls. And as you listen throughout the year, you will hear certain calls more than others in each season.

During late winter there is a particularly obvious change in the Chickadee's vocal habits, for the males start to give their Fee-bee-song — a clear, two-note phrase. This call becomes frequent at the same time that the Chickadees' small winter flocks begin to break up. When breeding males start to define their territories, the Fee-bee-song becomes even more prominent, for two or more males may be giving it at the same time, seeming to answer each other.

The period just before egg-laying and incubation is marked by the Teeship-call of the female. She gives this as she does the Wing-quiver display and follows her mate through the woods, possibly even being fed by him. Later, you are likely to hear several variable calls such as the Tseedeleedeet-call and the Che-beche-call; they are used by males and females when their territories are intruded upon by other Chickadees.

Near the nest in midsummer you may hear calls associated with breeding. One is a quiet version of the Fee-bee-song used by the adults as they approach the nest. The other is the Teeship-call again, this time used by the young in their fledgling stage as they beg for food from the parents.

In late summer and fall, the flocks that will stay together

through the winter begin to form and roam about their territories feeding. Two calls are commonly used by the birds at this time, the Tseet-call and the Chickadeedee-call. They are believed to function as contact-calls — sounds that enable the flock to stay together even though they may not be in visual contact.

BEHAVIOR CALENDAR

	TERRITORY	COURTSHIP	NEST-BUILDING	BREEDING	PLUMAGE (MOLTS)	SEASONAL MOVEMENT	SOCIAL BEHAVIOR
JANUARY							
FEBRUARY							
MARCH							
APRIL							
MAY							
JUNE							
JULY							
AUGUST							
SEPTEMBER							
OCTOBER							
NOVEMBER							
DECEMBER							

DISPLAY GUIDE

Visual Displays

Head-Forward

Male or Female *Sp Su F W*

The body is held horizontal, and the head is thrust foward with bill often gaping. The crown is raised and contour feathers ruffled.

CALL Chebeche-call and others

CONTEXT Given during close encounters between competing birds: a threat of attack that usually causes the other bird to fly away. *See* Social Behavior

Wing-Quiver

Male or Female *Sp Su*

Wings are lowered and opened slightly and then rapidly quivered. Birds are often crouched while displaying.

CALL Teeship-call

CONTEXT Done by fledglings when begging for food; done by female toward male during incubation and the first half of the nestling phase; and done by both male and female before and during mating. *See* Courtship, Breeding

Auditory Displays

Fee-Bee-Song

Male *Sp Su F W*

A clearly whistled two-note phrase, the first note usually about a whole tone higher than the second; often "answered" by another bird

feeeebee. feeeebee.

CONTEXT A loud version given during territorial skirmishes; a soft version given during mate-feeding. *See* Nonbreeding Territory, Breeding

Chickadeedee-Call

Male or Female *Sp Su F W*

chickadeedee. chickadadeedeedee.

A call almost exactly like the written sound, with the emphasis usually on the first two syllables

CONTEXT Given especially in late summer and winter when the birds are in flocks: given by a bird that has become slightly separated from the flock, or given after a disturbance has dispersed the flock; has the effect of bringing the flock back together. *See* Social Behavior

Tseet-Call

Male or Female *Sp Su F W*

tseeet. tseeet.

A high, short, single sound; very quiet; hard to hear at first

CONTEXT Given mostly by birds in a flock as they quietly feed; may help the flock keep in aural contact. *See* Social Behavior

Dee-dee-Call

Male or Female *Sp Su F W*

deedeedeedee

A repeated *dee* sound with a definite scolding quality

CONTEXT Given in conflict situations and often followed by other aggressive actions such as a chase; used especially in territorial skirmishes. *See* Territory

Chebeche-Call

Male or Female *Sp Su F W*

A fast call of three or more syllables with the emphasis on the last syllable

chebeche or *chebechebechebeche*

CONTEXT Given in conflict situations by the more dominant bird; usually has the result of making the other bird fly away. *See* Territory, Social Behavior

Tseedeleedeet-Call

Male or Female *Sp Su F W*

A sputtering call much like the written sound, with the accent on the last syllable

tseedeleedeet

CONTEXT Given during skirmishes and chases; one of the more common of many variable calls given in such circumstances. *See* Territory

See-See-Call

Male or Female *Sp Su F W*

A rapidly repeated series of short, extremely high whistles, almost like the squeaking of a shrew or mouse

seee seee seee seee

CONTEXT Given especially when predators or danger are spotted; often causes surrounding birds to freeze momentarily, or at least to be very alert

Teeship-Call

Male or Female *Sp Su*

A high call accompanying Wing-quiver; much like its written sound, with emphasis on the first syllable

teeship, teeship, teeship,

CONTEXT Given by female during mate-feeding and by young in fledgling phase. *See* Courtship, Breeding

BEHAVIOR DESCRIPTIONS

Territory

Black-Capped Chickadees are unusual in terms of territory. Like some other birds, they hold both breeding and nonbreeding territories, but unlike any of our other common birds, their nonbreeding territory is occupied and defended by a flock and not by an individual bird or mated pair. These flocks are highly structured and have predictable patterns of movement. (*See* Social Behavior.)

Breeding Territory

Type: Mating, nesting, feeding
Size: 10 acres; variable
Main behavior: Fee-bee-song, aerial chases, and various calls
Duration of defense: March through July

Hearing the clear, two-note Fee-bee-song of the male Chickadee in late winter or early spring is a sign that behavior patterns associated with breeding are beginning to take place. Male Chickadees form territory by becoming increasingly intolerant of the other members of their winter flock.

Territory formation is a gradual process, with a pair first separating from the winter flock, then remaining in a given area, and then finally defending that area from the intrusion of others. This all occurs over a period of a few months, the actual defending of an area lasting only a few weeks around the time of nest-building. The defended area generally decreases in size during the incubation and nestling phases and disappears altogether during the fledgling phase.

Territorial skirmishes take a fairly typical form. First there is the Fee-bee-song by one or more males, often in a type of vocal duel where they alternate giving the song. The song is given only when the birds are ten yards or more apart. As they get closer, the song stops, and a variety of aggressive calls are given: Chebeche-call, Tseedeleedeet-call, and Dee-dee-call. These are accompanied by short chases among trees or shrubs. Often one or both birds will appear to feed throughout these interactions. The skirmish ends when one of the birds leaves.

These interactions are usually brief early in the season, but at the time of nest-excavation in April through June they may last for forty-five minutes or more. They generally occur at the borders of a territory or near the nest site.

The male does not use song to advertise his territory, but generally sings only when he comes across another male there. The size of breeding territories has been estimated at ten acres. In July, when the young disperse, the boundaries of the breeding territory are no longer defended.

Nonbreeding Territory

Type: Feeding
Size: 20 acres
Main behavior: Dee-dee-call, Chebeche-call, Chickadeedee-call, aerial chases
Duration of defense: August through February

In late summer and early fall many Black-Capped Chickadees begin to form small flocks. (*See* Social Behavior.) These become stable in membership and define territories in which they remain until the beginning of the breeding season. A nonbreeding territory is often formed around the territory of a pair that has bred the previous season, and is usually about twice the size of a breeding territory.

The flock circulates around the territory, stopping to feed at certain productive spots. If a neighboring flock trespasses, aggressive interactions follow with chases and several characteristic calls: the Dee-dee-call, Chebeche-call, and Chickadeedee-call. These interactions are common and very obvious, for the feeding flock is normally quiet as it moves along, giving only the Tseet-call to stay in contact. The added aggressive calls are a sign that there is a boundary conflict, and then, when the flock is quiet again, it means the conflict has stopped.

In winter, when it is easy to keep track of a flock, it is

interesting to follow it about its territory. While doing this you are likely to witness one of these flock encounters from start to finish. You will also get an idea of the dimensions of the flock's territory. The nonbreeding territory breaks up in late winter as the males give the Fee-bee-song and become less tolerant of other flock members.

Courtship

Main behavior: The pair separate out from the winter flock and move about the territory together; the female Wing-quivers; the pair does bill touching.
Duration: March through June

There is no obvious stage of courtship in Chickadees, but you will notice the breakup of the winter flocks as pairs of birds separate out and begin feeding on their own. These pairs become intolerant of the presence of other Chickadees: if others come near, the pair will incite brief skirmishes with scolding calls and chases.

About the time of the excavation of the nest hole, the female may start displaying with the Wing-quiver as she follows closely behind the male. Accompanying this display is the Teeship-call. Every so often the male will approach her as she displays, and the birds will touch beaks. Actual feeding may or may not take place. This mate-feeding is very similar to parent-fledgling feeding, but that takes place later in the summer. Mate-feeding is a good indication of the pair's breeding stage.

Mating also takes place at about this time, but it is not commonly seen. During mating, the female and/or the male begin to Wing-quiver. The female crouches down, the male steps onto her back, and copulation occurs.

Nest-Building

Placement: In the sides of partially rotted trees, 4–15 feet high
Size: Nest-hole entrance about 1 inch in diameter
Materials: Excavated into wood, lined with soft fibers

Chickadee nests are excavated in soft, partially rotted wood. Often birches are used, for the bark remains intact while the inner wood becomes soft with rot. The nests may be excavated in the side of a trunk or actually down the end of a broken-off limb. They are generally four to fifteen feet off the ground.

The birds have a marvelous habit of carrying the excavated wood chips away from the nest and dropping them from a perch nearby. Seeing the chips drop from their bills is a giveaway of the nest's location. The pair often work in close association, alternately carrying wood chips from the nest.

It is common that nests are started or even completed and then not used, so even if you find birds excavating a nest, be prepared to have them abandon it for a new site.

Locating the Nest

WHERE TO LOOK In wooded areas in the suburbs or the country
WHEN TO LOOK Nest excavation from April through June
BEHAVIORAL CLUES TO NEST LOCATION:

1. Follow a bird after it drops wood chips from a perch.
2. Small wood chips littering the ground mean the nest is nearby. (I locate at least a third of the nests I find this way.)
3. Be aware of territorial skirmishes and watch for excavation or feeding of young in that area.

Breeding

Eggs: 6
Incubation: 12 days, by the female only
Nestling phase: 16 days
Fledgling phase: 1–2 weeks
Broods: 1–2

Egg-Laying and Incubation

During this stage the female is often fed by the male. He approaches the nest with food and gives the soft version of the Fee-bee-song. The female then leaves the nest and joins him. As she receives food, she Wing-quivers and gives the Teeship-call. Sometimes the female initiates this process by leaving the nest and calling the soft Fee-bee-song, and then if the male approaches she is fed by him; otherwise she feeds on her own. These feedings occur about once every half hour.

The eggs are laid one each day until the clutch is complete; the average clutch size is six eggs. The female does all the incubation, which lasts about twelve days.

Nestling Phase

The female broods the young for the first few days after they have hatched. The male continues to bring food to the nest, signaling with the soft Fee-bee-song each time he approaches. The female may still leave the nest and Wing-quiver in front of him, but he now gives the food to the young in the nest. When brooding time decreases, both parents join equally in bringing food to the young. The male continues to give the soft Fee-bee-song all through this phase, whereas the female is generally silent as she approaches the nest. The nestling phase lasts sixteen days before the young are able to leave the nest.

Fledgling Phase

Once the young leave the nest, they start to give a distinctive call much like the call of the female when she receives food from the male. This is the Teeship-call. The young will be recognized by their use of this call and by their habit of following the parents. After about ten days the parents no longer feed the young and may even be aggressive toward them. The young disperse in the following few weeks and actively engage in skirmishes with each other that involve much calling, and often imperfect renditions of the Fee-bee-song.

Plumage

Black-Capped Chickadees go through one complete molt per year in July and August. The birds' appearance is not changed significantly by the molt.

As there are no marked differences in the plumage of the male and the female, behavior must be used to distinguish the sexes, and even this can be of help only in certain seasons. In spring the male gives the Fee-bee-call and feeds the female before and during incubation. Only the female incubates. During fall and most of winter, male and female behavior is very similar.

Seasonal Movement

From careful behavioral studies it is clear that many Chickadees remain on or near their breeding grounds throughout the year. The fact that Chickadees seem more plentiful in winter is often due to their habit of staying in small, fairly conspicuous flocks during that time.

But there is also evidence that some Chickadees migrate, for there are large rises in local populations during fall and spring. In much of this movement the birds may be drifting to new areas rather than undergoing real north-south migration, but then, banding reports also show significant north-south movement for some individuals.

Social Behavior

In late summer after the young have dispersed, Chickadees gather into small flocks that remain together until the start of the next breeding season. This pattern seems to be true for the majority of Chickadees but may not be the case with them all. A flock usually forms around a dominant pair that has just finished a successful brood. The flock contains six to ten birds, some juveniles, some paired adults, and some single adults. It establishes a feeding territory which it defends against other neighboring flocks.

The flocks are easy to recognize, for they are small bands of birds moving slowly through the woods and continually giving the Tseet-call to keep in aural contact. You may see some short chases between members of a flock. This is probably the result of one member's expression of its dominance over another, for the flock has a linear hierarchy, with the two members of the main pair each dominating all others of their respective sexes. These expressions of dominance are best seen in crowded feeding conditions, such as those around a backyard feeder. If you watch closely you will see certain birds approach immediately, while others will wait until the more dominant ones are through.

Chickadee flocks are often joined by other species as they move about their territory feeding. Throughout the flocking period (August through February), other species frequently seen with the Chickadees include Downy Woodpeckers, Tufted Titmice, Kinglets, Brown Creepers, and White-Breasted Nuthatches. These are called "mixed flocks."

During fall migration, the Chickadees are also joined by many species of Warblers. The Warblers migrate at night and land early in the morning to feed. At this time they all give soft contact-calls to stay together, but by midmorning most of them have joined resident Chickadees and become silent, relying on the contact-calls of Chickadees to keep them together. It may be that the migrating birds are taking advantage of the resident Chickadees' knowledge of feeding sites.

An interesting feature of flock movement is that when Chickadees cross an open space, they usually fly only one or two at a time. Because of this it is possible to count the number of birds in a given flock, and this may help you to distinguish one flock from another in a given area.

Feeder Behavior

Chickadees are likely to be the first and most regular visitors to any suburban or country feeder. Their favorite food is sunflower seeds, and they are also attracted to suet. They will continue to come to feeders through summer if they can find a suitable nest hole nearby. In winter, Chickadees generally stay in small flocks that are fixed in membership and that defend a territory. A feeder at the edge of two territories will be visited by both flocks, but when the flocks meet, there will be some scolding between members. Within each flock there is a hierarchy, and the most dominant birds usually feed first and at the best spots. A bird may express dominance by taking the perch of another bird or by giving a call, which makes the other bird fly off. Subordinate birds may wait until a dominant bird has left and then go feed. See the sections on Territory and Social Behavior.

Most common displays: Head-forward, Dee-dee-call, Chebeche-call, and Tseedeleedeet-call all occur in close interactions and are usually given by the dominant bird. See the Display Guide.

Other behavior: Once the breeding phase starts, winter flocks break up and you will have fewer Chickadees at your feeder. If one or two pairs remain in the area to breed, you may see the female do Wing-quiver as she is fed by the male in courtship, and later you may see the young do Wing-quiver as they are fed by the parents. See the sections on Courtship and Breeding (Fledgling Phase) and the Display Guide.

House Wren / *Troglodytes aedon*

IT IS NEVER HARD TO TELL WHEN THE HOUSE WREN HAS ARRIVED on its breeding ground, for the male immediately starts in with his rich, bubbling warble. This is the time to start observing the Wren, because the early territorial and courtship activities of this bird are its most entertaining features.

Once you hear a male singing, locate it and watch its movements. Its territory is small, so it is usually easy to keep track of the bird. There are two parts to territory formation you should watch for: loud singing from a few prominent perches, and the placing of short, straight twigs in prospective nest holes. With fifteen to thirty minutes of watching, you are bound to have discovered one nest hole and will have a good sense of the territory. Returning on other days may reveal other nest holes, and you may see territorial disputes with neighboring males.

If you have spent some time watching the male's behavior before the arrival of the female, then you will easily detect the differences in his Song when he sees a female near his territory. It becomes very high and squeaky, and the bird vibrates his wings when he gives it. You will see the male seemingly "leading" the female to his various nest sites, and interesting interactions may occur as the female either does or doesn't enter the nest hole. Her acceptance of it will be clear when she starts to add a lining of soft grasses to the nest.

The rest of the breeding period is also enjoyable to watch, as the male often brings food to the female while she incubates the eggs. In most parts of the country the pair has a second brood, before which the male cleans out old nests and repeats his nest-building behavior along with resuming territorial singing.

House Wrens nest in natural tree cavities or old Woodpecker nests, but they seem to prefer nest boxes when available. Therefore, if a conservation area near you has nest boxes, watch for House Wrens using them. An alternative is to put up a nest box in your backyard, for this may attract a Wren.

BEHAVIOR CALENDAR

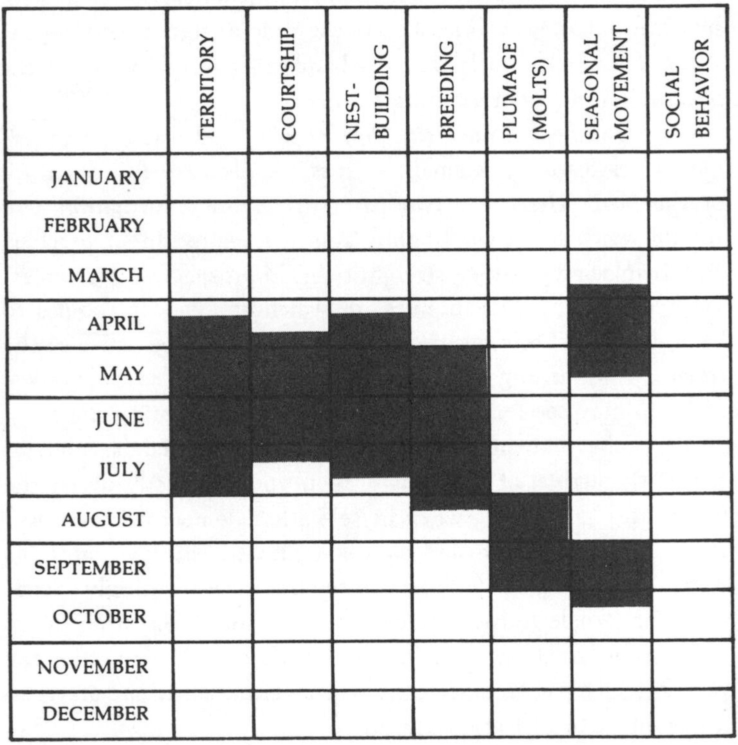

	TERRITORY	COURTSHIP	NEST-BUILDING	BREEDING	PLUMAGE (MOLTS)	SEASONAL MOVEMENT	SOCIAL BEHAVIOR
JANUARY							
FEBRUARY							
MARCH							
APRIL							
MAY							
JUNE							
JULY							
AUGUST							
SEPTEMBER							
OCTOBER							
NOVEMBER							
DECEMBER							

DISPLAY GUIDE

Visual Displays

Tail-Up

Male or Female *Sp Su F*

The House Wren's tail is normally in a down position but is raised more and more toward the vertical with increasing excitement or disturbance.

CALL None or Churr-call

CONTEXT Given in situations of danger or conflict, as in territorial disputes or when there is danger to the young

Wing-Quiver

Male or Female *Sp Su*

Wings are held out to the sides and quivered or fluttered rapidly.

CALL Song or Shrill-song

CONTEXT Occurs during close contact between mates, such as during courtship and during interactions at the nest. *See* Courtship

Tail-Spread

Male or Female *Sp Su*

The bird is horizontal, with head forward, tail lowered and spread, wings partially spread, and back feathers erect.

CONTEXT Occurs mostly at the nest site and is directed at predators or intruding males. *See* Territory

Flutter-Flight

Male *Sp Su*

The bird flies slowly with conspicuously heavy wingbeats.

CONTEXT Given by male during courtship as he flies to his nest site in front of his mate. It may help to attract her attention to the site. *See* Courtship

Auditory Displays

Song

Male *Sp Su*

A rich descending warble lasting two to three seconds; repeated many times with a short pause between

CONTEXT Given during territory formation while the bird is perched on one of its prominent song posts. *See* Territory

Shrill-Song

Male *Sp Su*

Similar to the warbling Song but with many high-pitched squeaks added, making it sound more harsh

CONTEXT Given by male when female first enters his territory, sometimes accompanied by slight Wing-quivering. *See* Courtship

Quiet-Song

Male *Sp Su*

Similar to the Song, but softer and shorter

CONTEXT Given by male at the nest, and seems to help the mates coordinate their care of the young

Buzz-Call

Male or Female *Sp Su*

A series of short buzzes *bzzz, bzzz, bzzz,*

CONTEXT Directed at predators or possible danger near the nest

Churr-Call

Male or Female *Sp Su*

A short burst of sound like a rattle being shaken

CONTEXT Given by male or female during aggressive encounters, or in relation to possible danger

BEHAVIOR DESCRIPTIONS

Territory

Type: Mating, nesting, feeding
Size: ½–¾ acre
Main behavior: Singing from perches, defending borders, putting stick foundations in prospective nest holes
Duration of defense: From arrival on breeding ground until fledgling phase of last brood

Territory formation and defense is an important aspect of House Wren behavior, and easy and enjoyable to watch. The males arrive first in spring and immediately begin to establish territory in two main ways: by repeatedly giving Song from exposed perches, and by laying stick foundations in possible nest sites.

Upon arrival, the male will start to use two to three prominent perches for delivering his territorial Song. Often neighboring males will alternate Songs. The Song may sound a little shaky in the first few days, and this is the case with both first-year males and older adults, but it soon becomes the rich warble that is easily recognized and heard from quite a distance. You should listen closely to the Song at this time, for in the following weeks there are two subtle variations that will alert you to other aspects of the male's behavior. When the female arrives, the Song takes on a harsh and squeaky quality, and when the young are in the nest, it is given in a quieter and often shorter version. *See* Courtship and Breeding.

Along with Song, the territorial male also starts to claim nest sites within his territory. House Wrens nest in tree holes or nest boxes. These are first cleared out of nesting material from past seasons, and then the male makes repeated trips to the nest, each time with a stiff twig about four inches long. These form the nest's foundation. (*See* Nesting.) From one to seven nests may be started in this way. Nest-building and singing occur, in varying proportions with each individual, throughout the period of territory formation.

A third part of House Wren territory formation is defense against intruding males. The territory is generally well defended, but males may leave their own territory to explore for new nest sites. When leaving, they are quiet and stay low within shrubbery. They may actually try to take the sticks out of a neighbor's nest area and claim it for themselves. In these cases the territorial male will fly at the intruder and land with the Tail-spread display, which seems to function as a threat, and either the intruder flies off or the two birds actually fight.

Territorial activities for the second brood are exactly like those of the first brood, so there are two good chances to see the whole process occur, once in spring and once in midsummer.

Courtship

Main behavior: Male singing, female inspecting nest sites
Duration: A few days

The courtship activities of the House Wren are most easily discerned if one has spent at least some time watching male territorial behavior. Knowing the boundaries of the territories and the location of the nest sites will help you know when a female has entered a male's territory. When a female first arrives, she may be attacked by the male as if she were another intruding male. But generally the female persists in the territory and responds with the Churr-call. Once the male recognizes that the new bird is a female, he sings the Shrill-song, which is accompanied by trembling movements of his wings and tail. He will then do Flutter-flight before the female and arrive at one of his nest sites. He may go in and out of the entrance and then sit nearby giving more of the Shrill-song. The female may then enter the box. This whole sequence may be repeated at the same box or at other nest sites.

Once the female has begun to bring material into one of that male's nest sites, that is a signal that she has decided to breed there. The male will often fly after the female as she moves about the territory. It is unusual to see copulation in House Wrens, but when it occurs it is initiated by the female as she gives a high, squeaky call.

Females may inspect nest sites on their own without the male's being near, and in such cases their actions will be almost identical to those of intruding males. In fact, a territorial male may mistake an intruding male for a female inspecting a nest site, and display to him with the Shrill-song rather than chasing him off.

Nest-Building

Placement: In tree holes or nest boxes, 4–30 feet above ground
Size: Inside diameter about 2 inches
Materials: Base formed of twigs; lining formed of grasses, feathers, spider egg cases

Nest-building by House Wrens seems to be intimately tied to both territory formation and courtship. This is unusual among our common birds, most of which go through territorial and pair formation before they even start to look for a nest site. With the House Wren, nest-building starts soon after the male arrives on the breeding ground. Much of his time is spent cleaning out old material in the nest sites and filling them with short, straight twigs and some spider egg cases. He may claim from one to seven nests, but they will not all be worked on equally — usually there are one or two he puts the most effort into. During this stage before the female arrives, the male may also try to claim a neighbor's nest site by pulling out its nesting materials and adding his own. After the female arrives, she seems to signal her acceptance of a mate by putting the lining into one of his nest foundations. Only the female puts the lining in the nest. If she chooses a nest that is not well built up, she may have to add more twigs. Before the female arrives the male works steadily on his nests, making endless minor alterations. But the female's work on the nest is restricted to about four days, and during the first two days she does the bulk of the work, averaging about 150 trips to the nest each of those days.

Tree-hole sites are much in demand in many areas and the House Wren is a strong competitor. Its main competition is with House Sparrows, and in some areas with Bluebirds, Tree Swallows, Red Squirrels, Chipmunks, Deer Mice, and Paper Wasps. The Wren is about on equal ground with its bird competitors, but cannot compete with the mammals. The male Wren cleans out all nest sites in his territory for both the first and second broods, and these may include the nests and eggs of other birds, in which case their broods are destroyed.

Locating the Nest

WHERE TO LOOK In nest boxes or in old trees where nest holes have already been excavated; in holes in trees generally located near the border of a woods

WHEN TO LOOK In spring and midsummer, especially during the territory formation of the male

BEHAVIORAL CLUES TO NEST LOCATION:

1. The male works regularly on the nest during territory formation for each brood, and as he carries twigs you can follow him to the nest.

2. If you see a Wren with grasses or feathers in its beak, it is probably the female putting the lining in her first or second brood nest.

3. During the nestling phase, the parents are conspicuous as they keep returning with food to the nest.

4. Any time you find a male singing, a nest cannot be far away, since the territories are only one-half to three-quarters of an acre in size.

Breeding

Eggs: 5–6
Incubation: 12–15 days, done by female only
Nestling phase: 16–17 days
Fledgling phase: 14 days
Broods: 1–2

Egg-Laying and Incubation

Between the times when the nest is completed and when the first egg is laid, there is a period of about two days. Generally five to six eggs are laid, one each day. Full-time incubation starts after the last egg is laid and continues for about thirteen days. The incubation rhythm of the female is around twelve minutes on the nest for every eight minutes off. She may leave the nest on her own, or the male may arrive at the nest and sing the Quiet-song, which is generally followed by the female's leaving to feed. While the female is off the nest the male stays nearby. The female does all the incubation, and roosts in the nest hole at night, as she has done since the lining was added. Once the female has added the lining to the nest, the male rarely enters it until the young have hatched.

Nestling Phase

The young generally hatch all on the same day. They are brooded by the female for about half of each day for the first three days of their nestling life. Both parents feed the young, although at first the male will not enter the nest, but passes food to the female, who then feeds it to the young. He often arrives at the nest giving the Quiet-song, and the female sticks her head out of the nest. The male Wing-quivers as he passes the food on to her, as they both do whenever they meet during this phase. She takes the food in, and he goes to a perch, sings, and then flies off. This is a frequently repeated pattern. Later in the nestling phase, the male will enter the nest when the female is not there and both the male and the female may be seen carrying fecal sacs from the nest. After about sixteen to seventeen days the young are ready to leave the nest.

Fledgling Phase

After the young leave the nest they continue to be fed by the parents. They are conspicuous at this time, for they stay together, and the attendant parents are quick to give the Churr-call when there is any possible danger.

After one to two weeks the female may leave the young to start a second brood. The male then feeds the young, but must also maintain his territory and prepare new nest sites.

Only about half the time do the Wrens have the same mate for the second brood.

Plumage

The House Wren goes through one molt per year in late summer, and the appearance of the birds does not change significantly with the molt. As male and female are alike in plumage, behavior clues must be used to distinguish the sexes. The male is the only one to sing and is usually the one who lays the stick foundations in the nest holes. The female is the only one to put the softer lining in the nest, and does all of the incubation.

Seasonal Movement

The House Wren is clearly a migratory bird. In spring the males tend to precede the females north by a week or more. In late summer, when breeding is finished, the Wrens are more secretive and stay within the brushy areas of woods or pasture borders. After they fly south they continue their secretive habits, rarely singing, and remaining inconspicuous, at least until just before their northern migration in spring.

Mockingbird / *Mimus polyglottos*

IT IS HARD FOR A BEHAVIOR-WATCHER TO THINK OF MOCKINGBIRDS and not also think of territoriality, for this is undoubtedly the most prominent aspect of this bird's behavior. Not only are its territories small, sharply defined, and aggressively defended, but they are also formed twice each year — once in spring for breeding and again in fall to protect a winter food source. Add to this the fact that the birds are partial to living in urban areas, and you undoubtedly have the best of our common birds in which to observe territorial behavior.

The two periods of territory formation are easily recognized, for they are both marked by the Mockingbird's loud, imitative song. In spring only the male sings, but in fall either male or female may sing.

The fall and the spring territories are quite different, for the spring territory is centered on the nest, whereas the fall territory is centered on a source of food. The differences are reflected in the defensive behavior of the Mockingbird. In spring and summer it displays toward people, dogs, cats, snakes, large birds and any other animal that might prey upon the nest. In fall the bird displays toward Robins, Starlings, and Blue Jays — birds that compete for its source of food, such as berries and other fruits. In both cases, of course, the territory is defended against other Mockingbirds as well.

In spring the male sings loudly until the female arrives. Then there is a short period when he chases her, and either bird may give the Chjjj-call. Once you have heard this call and noticed that the male is quieter, you can be fairly sure that a female has arrived on the territory and that nest-building will be occurring in the next few days. Then, with the aid of the behavior descriptions, it is quite easy to follow the birds' behavior from egg-laying on through the fledgling phase.

BEHAVIOR CALENDAR

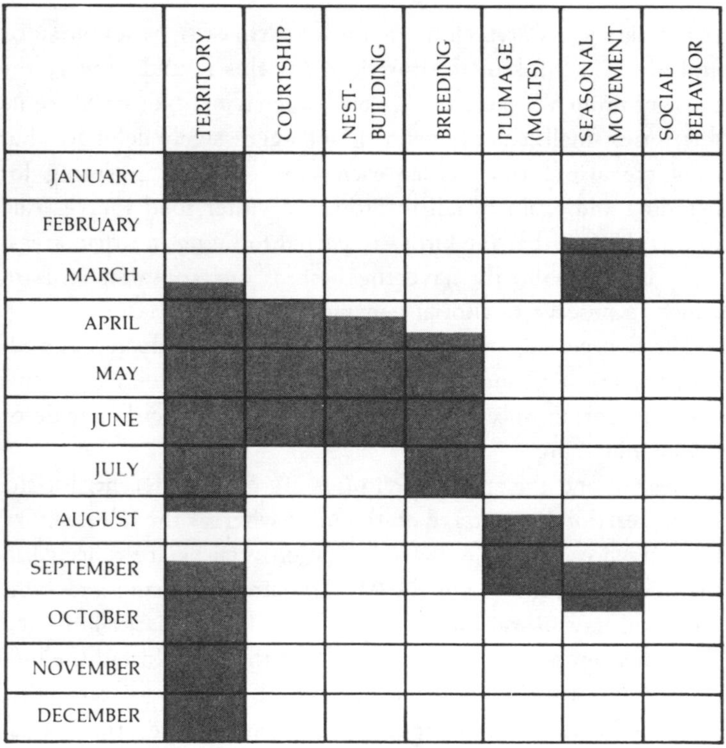

	TERRITORY	COURTSHIP	NEST-BUILDING	BREEDING	PLUMAGE (MOLTS)	SEASONAL MOVEMENT	SOCIAL BEHAVIOR
JANUARY							
FEBRUARY							
MARCH							
APRIL							
MAY							
JUNE							
JULY							
AUGUST							
SEPTEMBER							
OCTOBER							
NOVEMBER							
DECEMBER							

DISPLAY GUIDE

Visual Displays

Loop-Flight

Male *Sp*

Starting from an exposed perch, the bird flies up into the air, makes a short loop, and settles back onto the perch.

CONTEXT Given by males at the height of their spring territory formation, and may help to attract a mate. *See* Courtship

Mate-Chase

Male and Female *Sp*

A female is chased in flight by a male within his territory. Other males may join in the chase, in which case it may cross territorial borders.

CALL Chjjj-call

CONTEXT Given by a territorial male in spring when a female first comes near his territory; believed to be a part of courtship. *See* Courtship

Border-Dance

Male or Female *F*

Two birds on the ground, with tails and heads raised, slowly hop back and forth and from side to side.

CONTEXT Occurs near the end of fall territory formation at a border between two adjoining territories; may be a reaffirmation of territorial borders between known neighbors. *See* Nonbreeding Territory

Drooped-Wings

Male or Female *Sp Su F W*

While perched or on the ground, the bird lowers its wingtips well below the level of its tail.

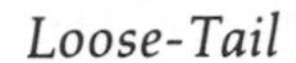

CONTEXT Given in the presence of other birds that are intruding upon the displayer's territory; often a precursor of more aggressive action, such as a direct attack. *See* Territory

Loose-Tail

Male or Female *Sp Su*

The bird spreads its tail slightly, making the white border feathers show, and then moves it loosely up and down and from side to side.

CONTEXT Given toward possible predators in the area of the nest, such as Crows, cats, squirrels, snakes or humans; often precedes attack. *See* Territory

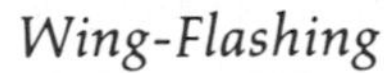

Wing-Flashing

Male or Female *Sp Su F*

The bird haltingly raises one or both wings, holds them open for a short while, and then lowers them. The display may be repeated.

CONTEXT Most often occurs when a bird is feeding on the ground; not known whether it has a communicative function

Auditory Displays

Song

Male or Female *Sp Su F*

witew witew witew, chicup chicup chicup, peer peer peer peer, etc. Only a sample.

A continual stream of other birds' calls, each call usually repeated three or more times

CONTEXT Given from exposed perches during

periods of territory formation and/or courtship; in spring used only by the male, but in fall used also by the female, since females often hold separate fall territories. *See* Territory, Courtship

Chjjj-Call

Male and Female *Sp*

A harsh, drawn-out, rasping call, often given while the bird is in flight

chjjjjj

CONTEXT Usually accompanies Mate-chase, and is one of the first signs of courtship or the arrival of a female in the area in spring. *See* Courtship

Chick-Chick-Call

Male or Female *F*

A repeated, short, harsh call given for long periods of time at the rate of about once every one to two seconds

chick. chick. chick.

CONTEXT Given from within cover and often heard at dusk at an early stage of fall territory formation. *See* Nonbreeding Territory

Ch'ch'chick-Call

Male or Female *F*

A short, rapid series of harsh notes

ch'ch'chick.
ch'ch'ch'chick.

CONTEXT Given by territory holders in fall when another bird that eats berries enters their territory; often causes neighboring Mockingbirds to give the same call. *See* Nonbreeding Territory

Chewk-Call

Male or Female F W

chewk. chewk. A single, loud, sharp sound

CONTEXT The only call of the Mockingbird in winter; given toward territorial intruders. *See* Nonbreeding Territory

BEHAVIOR DESCRIPTIONS

Territory

Mockingbirds have two types of territories: a breeding territory and a nonbreeding territory. The former is held from April until late summer, and the latter from September until February. Along with Chickadees, Mockingbirds are among the few common birds to hold a nonbreeding territory through the winter.

Breeding Territory

Type: Mating, nesting, feeding
Size: 1–2 acres
Main behavior: Song, chases, Drooped-wings, Loose-tail
Duration of defense: March through July

The start of territory formation by the male Mockingbird is obvious in spring, for it is marked by the male's Song. The Song is at first given low from within bushes and only intermittently through the day, but over the weeks it becomes louder and more frequent until it is almost continuous. The bird is made even more conspicuous by its habit of singing from one of several exposed perches within its territory.

If another male Mockingbird enters the territory, it is swiftly chased out with few or no calls. In a period of a month or so, most of the males in a given locality have settled territorial boundaries and rarely encroach on one another's areas. But still the song gets louder and even more persistent, and is accompanied by conspicuous, looping flights. This is generally considered to be part of courtship and an attempt to attract a mate.

The breeding and nonbreeding territories are defended against both birds and animals, but different intruders are more threatening in each of the seasons. The breeding territory is protected against possible enemies of the nest such as Crows, Blue Jays, snakes, squirrels, cats, and humans. Mockingbirds are often so bold in their defense that they will dive down upon the heads of humans who approach too near during the incubation and nestling phases, and effectively take over part of a backyard or park. The two displays most often accompanying these encounters are Drooped-wings and Loose-tail, both of which expose much of the white on the bird's wings and tail. If you see a Mockingbird in one of these postures, it is fairly certain that there is a disturbance nearby, even though you may not be able to see it. These displays frequently precede direct attacks on the intruder, taking the form of brief flights above the intruder, possibly followed by dives.

When the last brood has fledged, territorial boundaries are no longer defended.

Nonbreeding Territory

Type: Feeding
Size: 1–2 acres
Main behavior: Song, Ch'ch'chick-call, Chick-chick-call, Chewk-call, chases, Border-dance
Duration of defense: Displays from September through December; territory defended until about February

Mockingbird fall and winter territories are centered on a food source and seem to help assure that an individual bird will have enough food for the winter. The territories may be held by a single male, a single female, or a mated pair. As opposed to their habits in other seasons of the year, both male and female use Song in fall and winter and are similar in their behaviors. Winter territories may or may not be the same as summer territories. A pair that mated the previous summer may hold two separate territories or one common territory.

The first sign of renewed territorial behavior is the loud, widely spaced Chick-chick-call. It is an easy sound to discover, for it begins in early September when few birds are singing. These calls are believed to be given mostly by immature birds. In the second half of September, when most adults have completed their molt, they also begin claiming their winter territories, first with this same persistent, loud Chick-chick-call, and later with soft Song from within bushes. Further definition of territories is then announced by three different calls: the rapid Ch'ch'chick-call, the loud Chewk-call, and short, harsh bursts of Song.

From September until the beginning of October, each bird stakes out important perches from which it overlooks its territory, and there are frequent fights and calls between neighbors. By the middle of October territorial boundaries are generally fixed, and by watching an individual bird for ten to fifteen minutes you can usually get a fairly good idea of the size of its territory.

During fall, three types of intruders will be chased out of the territories. The most common intruders are other birds competing for the same foods (largely berries): Robins, Grackles, Starlings and Blue Jays. They are usually chased out by the territory holder diving at them and giving the Ch'ch'chick-call or the Chewk-call.

Another type of intruder is the lone Mockingbird, moving about and possibly looking for a territory. At first sight of this bird the territory holder generally moves to an exposed perch and gives the Chewk-call. If the area is well populated with Mockingbirds then other territory holders will respond to this call by also moving to high perches and calling. In this way the wandering bird will not be able to land in the area and must move on or be attacked.

A third type of intrusion comes from a group of Mockingbirds that for some reason do not have territories of their own. They typically persist in going after the food supplies of the territory holders. If this occurs, neighboring territory holders leave their territories temporarily and band together in scolding the intruders. When the "raid" is over the residents return to their own territories. These last two types of intrusion will be seen only in areas of high Mockingbird density.

By November, "raids" and wandering intruders have become few, and the residents are quieter and less conspicuous, often perching low and near to their most valued food source. During this month a new display may be seen along the common border of two territories. In the display, two (rarely three) birds meet on the ground, face to face, with tails up and bodies erect. In this position they slowly hop back and forth and from side to side, each bird a mirror image of the other. This may even continue down the territory line. The display, the Border-dance, ends either with a fight, with both birds' flying to a new spot and repeating the display, or with both birds' suddenly flying back into their own territories. It happens only at the borders of territories and is believed to help reaffirm the territorial borders of both birds.

From midwinter on into early spring, Mockingbirds are quiet, their only observable behavior being an occasional chase of another bird from their food source.

Courtship

Main behavior: Song, Chjjj-call, chases, Loop-flight
Duration: Variable

After the males have defined territories in the spring, they remain on them, and in order to get a mate they must wait until one comes into the area. They continue loud versions of their Song from exposed perches, possibly to help females locate them. Accompanying this are Loop-flights, where the bird circles up into the air and returns to its perch, or flies down to the ground and then back up to its perch, or flies slowly with a lot of white showing from perch to perch. The bird's singing continues day after day until a female arrives or the male gives up and leaves the territory. Some males without mates may keep singing well into midsummer.

When singing stops in an area where it has been fairly constant, that is a good indication that a female has arrived in the territory of that male. You can confirm this by seeing two Mockingbirds involved in a chase, and, during the chase, hearing one or both of them give the Chjjj-call. This is typical of the first interactions between male and female Mockingbirds. The male chases the female in a seemingly aggressive manner intermittently throughout the day — the repeated Chjjj-call should draw your attention to this stage of behavior. After a day or two of this activity, the chases stop, and both birds will be seen in close association around the territory. At this point the male sings softly and only a few times per day.

Two other displays may occur at this time, although they are not commonly observed. In one the male runs along a branch or similar object with his tail and wings spread and his head down. This is usually done near the female. In another he takes nesting material into his beak and flies into suitable nesting sites, sometimes followed by the female.

If, after the beginning of the breeding season, a male starts to sing loudly and constantly all day, there is a good chance that he has lost his mate for some reason. The males at any stage may be heard singing even in the middle of the night, especially if the moon is bright.

Nest-Building

Placement: In shrubs or trees 4–10 feet high
Size: Inside diameter 3 inches
Materials: Twigs, leaves, grape bark, fern rootlets, bits of plastic, string

Both male and female Mockingbirds participate in building the nest, and whenever a new clutch is started, a new nest is usually built. Often as many as five nests may be built by a single pair (some due to unsuccessful broods). Mockingbirds also build partial nests, for at the end of courtship the male may repeatedly pick up nesting material, fly to a suitable spot, and place it there. Often the female will follow him. Final nests are rarely built at these places; rather, the display probably serves to stimulate the female's nesting instincts and get her to build in the displaying male's territory.

The first layer of the nest is built of twigs, followed by finer twigs, some leaves and grape bark, and finally a lining of dark rootlets, which may be gathered from the base of old ferns.

Locating the Nest

WHERE TO LOOK In open meadow areas or in suburbs where shrubs are plentiful

WHEN TO LOOK April through June

BEHAVIORAL CLUES TO NEST LOCATION:

1. First locate the territorial boundaries by observing where the birds restrict their activity to.
2. Nest-building is not obvious and can be confused with the male's Nesting-display.
3. The best way to find a nest is to walk around within the territory and check the shrubs, gauging the birds' reactions to you as you do so. They will be visibly more disturbed as you approach the nest.

Breeding

Eggs: 4
Incubation: 12 days, all by female
Nestling phase: 12 days
Fledgling phase: Variable, 2–4 weeks
Broods: 1–3

Egg-Laying and Incubation

Four eggs is the average per clutch, and broods may be started anywhere from March through July, with the most occurring in

April through June. Only the female incubates, and the male may accompany her on her trips to get food or he may remain perched near the nest. During this stage of breeding the male gives only soft and intermittent Song.

Nestling Phase

The nestling phase lasts about 12 days, and during this time both parents feed the young to varying degrees. Often one adult remains near the nest while the other searches for food. For the first six days after hatching, the young will respond by begging if you tap on the nest. During the following six days, they respond to disturbances at the nest by cowering and being quiet. About the tenth day, the nestlings develop the *peeep* call, which they will use during their fledgling phase. Feces of the young are generally excreted over the rim of the nest rather than being carried out by the parents.

Fledgling Phase

Characteristic of the fledgling phase in Mockingbirds is the persistent, loud, and breathy *peeep* of the young. As with many birds, this helps the parents locate the young outside the nest. The fledgling phase can last as long as four weeks, during which time the parents may start a new brood. The fledgling Mockingbirds develop aggressive behavior in middle and late summer and can be seen chasing and scolding each other before they actually disperse.

Plumage

Mockingbirds go through one complete molt per year, usually in late summer. The plumage before and after the molt is not significantly different.

The plumage of males and females is identical in Mockingbirds, and so you must use behavioral clues to distinguish the sexes. One unusual aspect of Mockingbird behavior is that although only the male sings during the breeding season, both the male and female may sing at other times of the year. This correlates with their territorial habits, where only the male defends territory in summer, but both the male and the female may do so in winter. The female builds most of the nest and is the only one to incubate. During fall and winter, male and female are alike in behavior.

Seasonal Movement

The Mockingbird is continually extending its range into northern areas, so it would be hard to generalize about its seasonal movement patterns even if they were known. It is clear that many Mockingbirds remain in or near their breeding territories throughout the year. Males are more likely to remain in northern areas than are females. In late summer the young birds disperse, some moving only slight distances and others migrating south, then returning the next spring.

Feeder Behavior

Mockingbirds are not regular visitors to feeders since they prefer fruits and insects to seeds. A Mockingbird may establish a winter feeding territory that, by chance or intent, includes your bird feeder. The Mockingbird may then chase away other species from the territory, and thus from your feeder. If this occurs, move the feeder. See Territory.

Most common displays: The Chewk-call is the main call used in winter. It is given as the bird chases others out of its territory. See the Display Guide.

Other behavior: Mockingbirds have both summer and winter territories. Singing in spring and fall marks the start of territory formation. See Territory and Courtship.

Gray Catbird / *Dumetella carolinensis*

WHEN GRAY CATBIRDS ARRIVE ON THEIR BREEDING GROUNDS THEY soon begin to sing within their chosen territories. This is a good time for the behavior-watcher to locate two or three territories to which he or she can then return later to follow the behavior of specific pairs.

When the females first arrive on the breeding ground, you may see some fascinating displays. These are a challenge to observe, not only because they are uncommon, but also because the birds tend to inhabit dense, shrubby areas, where it is hard to keep them in sight. The best way to spot the displays is to watch for any continual chases, or for any situation in which there are more than two Catbirds. The displays include: a visual display where the body feathers are greatly fluffed, an auditory display of high, squeaky singing, and prolonged chases within limited areas. All these displays are believed to be associated with courtship, but there is still much more to be learned about them.

Catbird calls are easily distinguished. Some seem to be quite specific as to when and where they are used. For example, the Kwut-call is given by the parents whenever an observer is near the young during the nestling or fledgling stages. The Meow-call and Ratchet-call are given in situations of alarm. The Song is generally used to advertise territory, but the bird may also start singing in response to your approach.

The cooperation of the pair around the nest is one of the more interesting features of Catbird behavior. During the incubation phase the male is generally quiet but will approach the nest either to feed the female or to "guard" the nest while the female leaves to feed herself. Common displays at this time are Wing-flick and Raised-wing — either display is a good clue to the whereabouts of the nest.

BEHAVIOR CALENDAR

	TERRITORY	COURTSHIP	NEST-BUILDING	BREEDING	PLUMAGE (MOLTS)	SEASONAL MOVEMENT	SOCIAL BEHAVIOR
JANUARY							
FEBRUARY							
MARCH							
APRIL							
MAY							
JUNE							
JULY							
AUGUST							
SEPTEMBER							
OCTOBER							
NOVEMBER							
DECEMBER							

DISPLAY GUIDE

Visual Displays

Body-Fluff

Male or Female *Sp Su*

The bird fluffs all its breast and back feathers, puffing up to about the size of a softball. The tail is spread and often held down. In high-intensity versions, the wings may be raised.

CALL None, Soft-song, Meow-call

CONTEXT Given between competing Catbirds in territorial disputes, and given at predators; high-intensity version given at most serious predators, such as snakes. *See* Territory, Courtship

Raised-Wing

Male or Female *Sp Su*

The body is tilted horizontally and the wings are stiffly raised to the sides and held there for a moment.

CALL None, Meow-call, Kwut-call

CONTEXT Occurs especially near the nest, in response to your own approach, or that of a predator. *See* Territory

Wing-Flick

Male or Female *Sp Su*

The wings are repeatedly flicked in and out from the sides. The bird is usually horizontal.

CONTEXT Given mostly by the male as he approaches the nest; may be a signal that helps the pair coordinate their activities around the nest. *See* Breeding

Auditory Displays

Song

Male *Sp Su*

cheer, toowee, seechow, chuk, whooit, meeeow, etc. Only a sample.

A long burst of song with many different phrases in succession, each phrase usually done only once; often imitates the songs of other birds

CONTEXT Given from high conspicuous perches by the male during territory formation. *See* Territory, Courtship, Breeding

Soft-Song

Male *Sp Su*

A quiet version of Song, given only for a few minutes, not incessantly, as is Song

CONTEXT Given during aggressive encounters with predators

Meow-Call

Male or Female *Sp Su*

meeeow. meeeow.

A wheezy call like the *meow* of a cat; high, raspy, and variable

CONTEXT Given between Catbirds during aggressive encounters and given in the presence of predators. *See* Courtship, Breeding

Ratchet-Call

Male or Female *Sp Su*

tchtchtchtch

A grating short call, like the sound of a wooden ratchet

CONTEXT Given when the bird is suddenly alarmed, especially during nesting phases

Kwut-Call

Male or Female *Sp Su*

A short, throaty sound, much like its written description

kwut. kwut. kwut.

CONTEXT Given in moments of mild danger, especially during the fledgling phase. *See* Breeding

BEHAVIOR DESCRIPTIONS

Territory

Type: Nesting, breeding, feeding
Size: 1–3 acres
Main behavior: Song, chases, Body-fluff, Raised-wing
Duration of defense: From arrival of the male until young of last brood have fledged

Male Gray Catbirds start forming territory as soon as they arrive on their breeding grounds. This is done primarily by loud singing from a number of prominent perches around their territory. Disputes with neighboring Catbirds are generally settled with chases and singing by both members.

The birds stay within the territory throughout the breeding season, since feeding, nesting, and breeding all take place there. Their behavior exemplifies the popular conception of what territory is for animals. But after the female arrives and a nest site is chosen, the male does not continue to defend the borders of the territory as do some other species, such as Red-Winged Blackbirds or Mockingbirds, but spends its time defending a smaller area centered on the nest.

When the birds start a second brood, a new nest is built, and this shifts the focal point of the territory slightly. The male renews his singing at this time, and, as with the first brood, stops regular singing when incubation is under way.

The male Catbird defends the territory against birds of its own or certain other species, but the male and female both defend the area against nonavian intruders, the largest of which is undoubtedly you. Catbirds are very responsive to human intrusion onto their territories. When you enter a territory, the male may actually come up close to you and start loud singing, keeping you in view all the time. If you come near the nest he may give a number of responses: the Meow-call, the Body-fluff, or the Raised-wing display. I have even found some birds to give the Body-fluff display when I imitated the Meow-call.

As is also true of other birds, some males may define a territory but never attract a mate and so leave the area.

Courtship

Main behavior: Song, chases, Body-fluff
Duration: For the first few days after the female's arrival and possibly just before the second brood

Females migrate to the breeding grounds a week or more after the males, and by the time they arrive, the males have already established territories. Like many other birds, the male first chases the female about, just as he would any intruding male. But instead of fleeing, the female continues to stay in the territory. These chases are obvious, for they always remain within a small area and may last from a few minutes up to a half hour or more. The female generally keeps giving a few short, high notes while the chasing male gives Song fragments and the Meow-call.

Another display seen around the time of the female's arrival involves three or more birds. Most of the participating birds are in a Body-fluff posture, but also have their necks stretched and bills tilted up and slightly opened. In this position these birds typically hop along branches, often moving upward. A song may be heard that is like an extremely high and squeaky version of their normal Song. Often this pattern of behavior is seen along with the chases described above. The group behaviors remind one of the actions of Blue Jays in their early courtship flocks.

Nest-Building

Placement: In shrubs, vines, or small trees, 2–10 feet high
Size: Inside diameter, 3 inches
Materials: Sticks, small twigs, grape or cedar bark, and rootlets

Both the male and the female may be seen carrying nesting material at first, and a few preliminary nests may be partially constructed, but the final nest used for breeding is constructed solely by the female. Construction takes from five to eight days, and eggs are laid immediately after its completion. The materials used for the nest are almost identical to those used by Mockingbirds; in fact, the nests of the two species are so similar that it is impossible to tell them apart without seeing the birds in them. The only difference may be that Catbirds tend to use fewer man-made materials than do Mockingbirds.

Locating the Nest

WHERE TO LOOK In low dense growth at the edges of woods, usually bordering fields, streams, or lakes; best to get under the shrubs and look up

WHEN TO LOOK Starting about two weeks after the males first arrive and continuing throughout the summer

BEHAVIORAL CLUES TO NEST LOCATION:

1. Look within the territory.
2. Follow the male or female as they fly with food.
3. Check in areas where the Meow-call or Kwut-call is given, or where you have seen birds doing the Wing-flick.

Breeding

Eggs: 3–4
Incubation: 13 days, by female only
Nestling phase: Average 11 days
Fledgling phase: 2 weeks or slightly more
Broods: 1–2

Egg-Laying and Incubation

The female starts laying the eggs as soon as the nest is finished. One egg is laid each day until the clutch is complete. The female does all the incubation, generally staying on the eggs for about twenty-minute periods and leaving the nest for five-to-ten-minute intervals in between. The male and female coordinate their activities around the nest, starting with the period of egg-laying, and seldom leave the nest unattended for very long. To accomplish this, they use various visual and auditory displays.

At the start of egg-laying the male generally stops giving Song and becomes quiet and secretive as he moves about the territory feeding. Every so often he comes near the nest and may give soft fragments of Song. Upon hearing this, the female often leaves the nest to feed, possibly giving the Kwut-call as she flies off. The male then approaches the nest and stays within a few yards of it while the female is away. He often does the Wing-flick display during these periods of "nest guarding."

Sometimes the male approaches the nest with food and does the Wing-flick display; the female then leaves the nest and receives the food. In either case, when the female returns to the nest, the male leaves. This approximate coordination of activities continues through the incubation phase.

Nestling Phase

Both male and female take equal parts in caring for the young. For the first few days the female broods the newly hatched birds, and the male may bring food to her, which she in turn feeds to the young. For the first half of nestling life the fecal sacs are most often eaten by the parents, but during the second half they are usually carried away.

The nestling phase varies from seven to fifteen days, depending on when the young crawl out of the nest and remain perched on nearby branches.

Fledgling Phase

After a few days of perching near the nest being fed by their parents, the young start to move about more but still do not leave the territory. When the male assumes more of the responsibility of feeding the fledglings of the first brood, the female usually starts building a nest and laying eggs for the second brood.

During the fledgling phase you can use the parents' behavior to help you locate the young. If you are just in the general area of the young, the parents will approach you while giving the Meow-call. If you are very near the young, the parents become more secretive and change to giving the Kwut-call.

Plumage

Adult Gray Catbirds go through one complete molt per year in August, which creates little change in their appearance. Song stops during this period, and the birds may congregate in lowland areas. The main calls heard as you approach their haunts will be the Meow-call and the Kwut-call. The plumage of the male and female is identical, so behavior must be used to distinguish the sexes. The two main clues are that the male is the only one to give Song, and the female is the only one to incubate.

Seasonal Movement

After the last broods have fledged, Catbirds tend to move to areas of cover, where they remain for their molting period. In fall they migrate south, many going all the way to Central America. In spring the males start north about a week before the females, and upon arrival on their breeding grounds they begin to establish territories.

American Robin / *Turdus migratorius*

YOU WOULD THINK THAT OUT OF ALL THE BIRDS INCLUDED IN THIS guide, the American Robin would be the best known. In fact, there have been several major studies of its behavior and life history, but still there are some important mysteries surrounding its life. One of these concerns courtship. As yet, no one has discovered any displays regularly associated with courtship among Robins. If there were any, it is more than likely that they would have been seen by many people, considering how often Robins live near humans. If there aren't any, then exactly how do Robins form pairs? This has still to be solved.

Another enigma of the Robin's behavior is its Song. It is given by the male but does not seem to be associated with any particular aspect of the bird's life. In most birds Song is used to advertise territory, to attract a mate, or both, but in the Robin it is not strongly associated with either of these. The Robin sings most just before its young hatch, and this same frequency is repeated in each of its summer broods.

Robins have a variety of other auditory displays, most of which are used in aggressive interactions or during moments of alarm. You will find that the calls are extremely familiar, for we hear them all the time, but you will also probably be surprised at how many different ones the birds make. One unusual call heard in fall is high and thin, like that of the Cedar Waxwing, and is given by Robins as they fly in the evening to their large communal roosts. Roosting can be a surprising aspect of Robin behavior, for most people do not know that it occurs. Sometimes the roosts contain Starlings and Grackles as well.

Territoriality in Robins is also not as clear-cut as it is in many other species, for the territory is not always defended against other Robins, and many times an intruder prevails over an owner. Both of these instances are rare in other birds such as the Mockingbird and the Red-Winged Blackbird. Clearly, there is much more to observe and learn about this extremely common bird.

BEHAVIOR CALENDAR

	TERRITORY	COURTSHIP	NEST-BUILDING	BREEDING	PLUMAGE (MOLTS)	SEASONAL MOVEMENT	SOCIAL BEHAVIOR
JANUARY							
FEBRUARY							
MARCH							
APRIL							
MAY							
JUNE							
JULY							
AUGUST							
SEPTEMBER							
OCTOBER							
NOVEMBER							
DECEMBER							

DISPLAY GUIDE

Visual Displays

Tail-Flick

Male or Female *Sp Su F W*

The bird repeatedly flicks its tail sharply.

CALL Tuk-tuk-call

CONTEXT Given in situations of possible danger. *See* Breeding

Tail-Lift

Male or Female *Sp Su*

The bird lowers its head and raises its tail up to a forty-five-degree angle. The display is usually directed at another nearby Robin.

CONTEXT A display that precedes a direct attack. *See* Territory

Wing-Droop

Male or Female *Sp Su*

Wingtips are lowered so that they droop below the level of the tail. Breast feathers may be puffed out.

CONTEXT Occurs just before or after aggressive encounters. *See* Breeding

Auditory Displays

Song

Male *Sp Su F*

Three or more song phrases separated by a short space, then the whole pattern repeated after a pause

cheeriup, cheerily, cheeriup,

CONTEXT Sung early and late in the day, especially just before the young hatch in either brood

Teek-Call

Male or Female *Sp Su F W*

teeeek. teeeek.

A short, shrill call, sometimes repeated a number of times

CONTEXT Usually given in situations of possible danger; often accompanies the Tail-flick. *See* Breeding

Tuk-Tuk-Call

Male or Female *Sp Su F W*

tuk tuk tuk tuk or *teeeek. tuk tuk tuk tuk*

A halting series of short, guttural notes, often following the Teek-call

CONTEXT Usually given in situations of possible danger or disturbance; may accompany Tail-flick. *See* Breeding

Teacheach-Call

Male or Female *Sp Su F W*

teacheacheacheach or *teacheacheacheach-oochoochooch*

A rapidly repeated series of harsh notes often increasing in volume

CONTEXT Occurs in a variety of situations

Eee-Call

Male or Female *F W*

eeee. eeee.

A high, thin whistle call like that of Cedar Waxwings

CONTEXT Given while the birds are in flight to their evening communal roosts in fall and winter. *See* Social Behavior

BEHAVIOR DESCRIPTIONS

Territory

Type: Nesting, some feeding
Size: About 1/3 acre
Main behavior: Chases, fights, scolding calls
Duration of defense: From arrival on breeding ground through all broods

The first sign that male Robins are beginning to outline territories is the breakdown of the large social flocks of the winter months. The males become less tolerant of other males' being near them and show this intolerance with a number of aggressive interactions: the "attack-run," where the bird aligns its body with a horizontal plane and runs toward another Robin; the Tail-lift, where the bird lowers its head and lifts its tail to the vertical while facing an opponent; the "crouch," given by a bird when approached too closely by another bird (this often precedes an attack); and "pushing," a more subtle form of aggression where one bird simply keeps taking short runs toward another bird and the other bird keeps moving slightly away from it, making its own short runs. Pushing is a very frequent occurrence on lawns and other feeding areas.

Along with these aggressive interactions, male Robins will begin to restrict the majority of their movements to a particular area of up to about an acre in size. This may broadly overlap the areas of other males. After the females arrive and join with their mates, these areas become smaller. The final size of the territory is about one-third of an acre, and the pair will spend most of their time in it as well as build their nest there. The area will be defended through all broods though sometimes territory is switched or moved for the second brood.

Both males and females will defend the territory against intruding Robins of either sex, but territoriality in Robins is far more fluid than in birds such as House Wrens or Red-Winged Blackbirds. The territorial borders shift and are not well defined.

Neighboring territories seem to overlap, and birds are not always dominant in fights in their own territories. Also, territories are not always defended against intruders, and aggressive interactions may even occur outside the territory, such as on a common feeding ground.

The territory is a place where mating, nesting, and most of the feeding take place. But Robins also freely leave their territories for food, nesting material, and even roosting during the breeding season.

Male Robins often return to the same area in successive seasons, so the territory may be nearly identical to that of the previous year, and if not, then it is generally not far away.

Courtship

Main behavior: No obvious behavior

The overt signs of courtship are limited to the repeated presence of two birds together on a territory. There are accounts of isolated instances of pair-formation displays but these have been so rarely seen that they are most likely not common occurrences. Even copulation is rarely seen in Robins, although clearly it is common. What you will see is that after males have gone through the first stages of territory formation, the females will arrive on the breeding ground and join with mates.

The females and males generally return to the same area for breeding year after year, but banding studies have shown that they do not necessarily breed with the same mates.

Nest-Building

Placement: On a horizontal limb or building structure, 5–30 feet high
Size: Inside diameter 4 inches
Materials: Grasses, a middle layer of mud, and a lining of grasses

Robins are one of our earliest birds to nest, and their first nests are often placed in evergreens for protection, since the leaves of deciduous trees may not have come out yet. The female does the majority of building, although the male may help bring materials to the nest site. It is fairly common to see Robins flying with wads of muddy grass in their beaks. Nesting materials may be gathered up to a quarter of a mile from the nest site. Another sign of nest-building is a line of mud across the breast of the female, for she forms the nest by sitting in it and pressing her breast against the edges. The nest takes two to six days to build, and may be started up to two weeks before the first eggs are laid.

The nests may be used for more than one brood, and first-brood nests may be built right on top of nests from the previous season.

Locating the Nest

WHERE TO LOOK In open suburban areas; in evergreens, in sturdy shrubs, or on horizontal man-made surfaces

WHEN TO LOOK As soon as the females have begun to arrive on their breeding grounds

BEHAVIORAL CLUES TO NEST LOCATION:

1. Look for a bird gathering or flying with muddy nest material.
2. Look for birds with muddy feathers across the breast.
3. Be aware of Robins that continue to be alarmed at your approach.

Breeding

Eggs: 3–4
Incubation: 12–14 days, by only the female
Nestling phase: Average of 13 days, varies
Fledgling phase: Up to 4 weeks
Broods: 2–3

Egg-Laying and Incubation

The eggs are laid one each day until the clutch is complete. The female does all the incubation and generally remains on the eggs for about fifty minutes out of each hour. The male stays in the vicinity of the nest and shows up quickly if the female gives any calls of alarm. Sometimes the male may feed the female on the nest, but generally she leaves the nest to feed herself. The male is not near the nest at night and may actually roost communally with other males. If the eggs are destroyed for some reason, then a new clutch may be laid within ten days.

Nestling Phase

The nestling phase of Robins varies from as short as nine days to as long as sixteen days. The majority of young stay in the nest for thirteen days. During that time they are fed by both parents in varying amounts, and the female broods them for the first few days of their nest life. Robins are conspicuous as they gather food on lawns, and if they don't eat it immediately, but fly off with it in their beaks, then you can be fairly sure that they have young in the nest.

Robins are quite protective of their nest sites, especially during the nestling phase. Nest predators such as Crows will be mobbed in an area where there are a number of Robin nests. The birds will dive at the intruder, give the Teek-call, and repeatedly snap their bills closed, making a clicking sound as they dart about.

As you approach a nest the Robins may do Wing-droop and be quiet but watchful, or they may do Tail-flick and continually give the Tuk-tuk-call. If startled they may give a loud Teek-call and fly off.

Fledgling Phase

It is common to see young Robins with their speckled breasts running after one of their parents and calling for food. The male does the majority of the feeding of fledglings, for the female has often left to start the nest-building and egg-laying for the second brood.

Plumage

Robins go through one complete molt per year in July and August. The male and female are similar in appearance, except that the male generally has a darker head and his breast is a richer red. Other than this, the only behavioral clues to distinguishing the sexes are that the male is the only one to sing and the female does all of the incubation.

Seasonal Movement

After breeding is finished, Robins join into large flocks that roost together by night and feed together by day. They also change their habits from feeding on lawns to feeding on ripening fruits from shrubs and trees. Local flocks migrate first and are followed by Robins in more northern areas. Northward migration also occurs in large flocks.

As with many of our other common birds, the migration patterns of Robins are not clear-cut. There are often large flocks of from fifty to one hundred birds that remain in cold northern areas throughout winter. Since snow cover prohibits them from feeding on the ground, they subsist primarily on berries, larger fruits, and some seeds.

Social Behavior

At various times of the year, Robins gather for the night in large communal roosts. These roosts are quite obvious in fall after breeding has stopped and before migration occurs. The birds can be seen in the late afternoon, flying overhead a few at a time. They have a special call used during this flight, and it is much like the call of the Cedar Waxwing — a high, thin *eeee*.

Robins generally remain in flocks through the winter, and it is the break-up of these flocks in spring that signals the start of the breeding season. But even during the breeding period, male Robins roost together at night and then return to the area of their nests during the day.

Starling / *Sturnus vulgaris*

THE STARLING IS UNDOUBTEDLY ONE OF THE LEAST LOVED BIRDS IN North America, for its aggressive claiming of nest holes often crowds out other species, and its bothersome population growth seems to have no clear end in sight. In these respects *Sturnus vulgaris* is very similar to *Homo sapiens*. In any case, the behavior-watcher can find something of interest in every animal species, and since Starlings are always available, they make great subjects.

Some of the most interesting behavior of Starlings occurs around prospective nest sites. The birds start showing interest in nesting cavities as early as fall. At this time males and females begin to visit nest holes and remain near them for at least part of each day. A common display of the birds at the nest site is Crowing, in which they tilt their heads up slightly, give a continuous chortle of calls, and fluff out their throat feathers.

By about mid-spring, males are more active in relating to females in the area. When a female flies by the nest, the male does vigorous Wing-waving. The female may then return and enter the nest hole. If the male follows, bringing in sticks or leaves, then the birds will probably remain together as a pair. After pairing, the two share in their daily activities — feeding, perching, and flying to and from the roost together. Before this the birds have been independent.

Throughout the year, many Starlings gather into huge communal roosts each night. There are very typical behavior patterns associated with these roosts. Usually a few hours before sunset the birds start to gather in the tops of certain trees or buildings and call noisily. From these preroosting spots, groups of birds fly off toward the primary roost, often gathering with other groups and forming huge flocks that are conspicuous in the afternoon sky. These flocks swarm above the roost site, and members continue to make sharp dives into the roost trees. The roost may contain anywhere from a few hundred to a quarter of a million birds.

BEHAVIOR CALENDAR

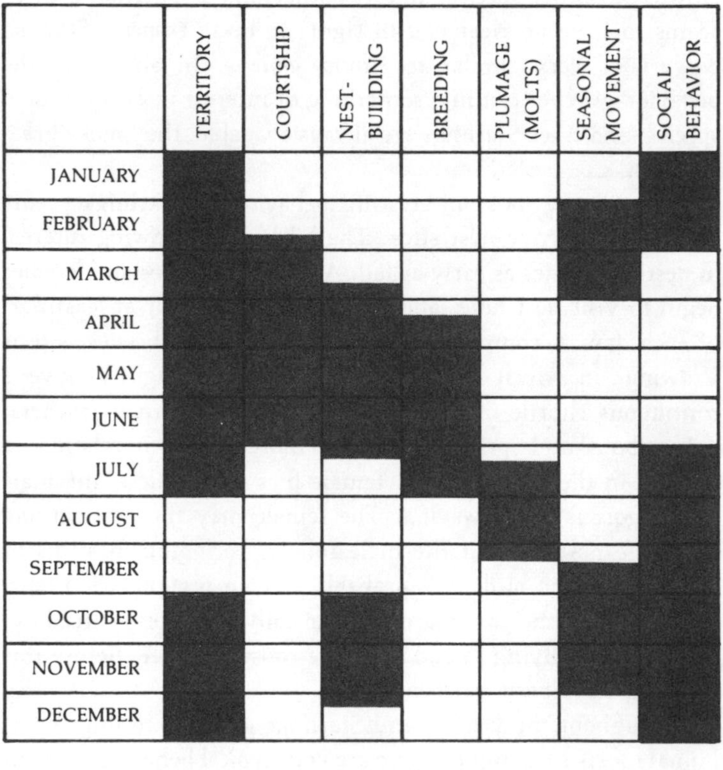

	TERRITORY	COURTSHIP	NEST-BUILDING	BREEDING	PLUMAGE (MOLTS)	SEASONAL MOVEMENT	SOCIAL BEHAVIOR
JANUARY							
FEBRUARY							
MARCH							
APRIL							
MAY							
JUNE							
JULY							
AUGUST							
SEPTEMBER							
OCTOBER							
NOVEMBER							
DECEMBER							

DISPLAY GUIDE

Visual Displays

Crowing

Male *F W Sp*

A perched bird tilts his bill up and fluffs out his throat feathers. His tail is held down in a vertical position, and his bill remains closed.

CALL Crowing-call

CONTEXT Generally done from a perch near a prospective nest hole; common during pre-egg-laying periods of all broods; also occurs at prerooosting sites and in connection with Sidling. *See* Territory

Wing-Wave

Male *Sp Su F*

While perched, the bird spreads his wings and moves them in a rotating manner.

CALL Squeal-call

CONTEXT Given primarily by the male near the nest hole, often when a female flies by; most common just before pairing in spring and on warm fall days. *See* Courtship

Fluffing

Male or Female *Sp Su F W*

The displaying bird faces another bird and puffs out all its feathers. The other bird may do the same.

CONTEXT Occurs during aggressive encounters. *See* Territory

Wing-Flick

Male or Female *Sp Su F W*

Only the tips of the wings are extended and rapidly flicked.

CALL None or Squeal-call

CONTEXT Given by both birds involved in aggressive encounters. *See* Territory

Sidling

Male *Sp Su F W*

The bird moves sideways along a branch toward its tip, forcing another bird farther out on the branch.

CONTEXT Occurs near nest holes and is a display of the nest-hole owner toward an intruder. *See* Territory

Bill-Wipe

Male or Female *Sp Su F W*

The bird rapidly and repeatedly wipes its bill on either side of a branch.

CONTEXT Done most often near the nest site by the nest owner when intruders arrive. *See* Territory

Auditory Displays

Song

Male *Sp Su F*

An extremely variable but somewhat melodious set of calls, often incorporating shrill squeals, squawks, and many imitations of other birds' calls and songs

CONTEXT Given near the nest and, unlike the Crowing-call, does not involve any action display

Crowing-Call

Male *F W Sp*

A continuous chortling call, variable and unmusical; given with bill closed

CONTEXT Accompanies the Crowing display. *See* Territory

Squeal-Call

Male *Sp Su F*

A vacillating, high-pitched squealing

CONTEXT Accompanies the Wing-wave display. *See* Courtship

Flock-Call

Male or Female *Su F*

A short call much like its written sound *djjjj*

CONTEXT Given by the large flocks of juveniles as they fly about looking for food

BEHAVIOR DESCRIPTIONS

Territory

Type: Nesting
Size: The immediate area of the nest site
Main behavior: Crowing, Bill-wipe, Fluffing, Wing-flick, and Sidling
Duration of defense: October through July to varying degrees

Territorial behavior in Starlings is limited to defense of the nest site, which is usually just a tree hole and nearby limb. There are a number of behavior patterns that will help you to recognize a bird that is claiming a nest site. First of all, the bird begins to spend part of each day perched near the site. While perched there, its most common activity is Crowing. The display becomes more intense whenever there is another male nearby, also Crowing at its own nest site.

A typical encounter starts with the arrival of a competitor. The owner then generally places himself between the competitor and the nest site. If the two birds are near each other, any of three subtle displays may be seen: Bill-wipe, Fluffing, or Wing-flick. Another more active display is Sidling. In this display the nest owner sidles along the branch toward the intruder, forcing it to move farther out on the branch and away from the nest. This continues until the intruder flies off or moves to a new perch.

Males defend nest holes against other males and try to entice females to them. In fall and winter, before pairing takes place, females may also defend nest sites. Generally they are aggressive only toward other females and will allow males to come near the nest and even enter it without conflict.

Defense of nest sites starts in early fall and increases until midwinter, when it diminishes slightly. In late winter it picks up and becomes more intense right up to spring, when breeding takes place.

Courtship

Main behavior: Wing-wave, Squeal-call, chasing, and birds remaining together during the day
Duration: A week or two before nest-building

The first and most obvious aspect of Starling courtship is Wing-wave, accompanied by the Squeal-call. In this display the male remains perched near his nest site, and whenever he sees another Starling fly overhead he starts to wave his wings in a circular motion and give a high-pitched squealing call. In most cases the bird in flight does not stop, but occasionally one will turn and approach the displaying male, and it is frequently a female. This seems to be a long-distance advertisement of the male for a mate.

When a female is near a male's nest site the male may pick up tree flowers or leaves in its beak and then repeatedly go in and out of the nest. This may go on for long periods with seemingly little effect, and the two birds may leave the nest area and return later with many of the same activities.

The main change in behavior that will tell you two Starlings have paired is that they will go through their daily activities together. Before this and earlier in the season, Starlings have stayed by the nest, fed, or gone to the night roost as individuals, but after pairing they do these activities together. Thus, in spring, when you begin to see Starlings flying in pairs, you can be fairly sure that broods will soon be started.

A chase flight of the female by the male has been repeatedly observed around the time of pairing and may be associated with pair formation. Such chases are common in the courtship of other species. Mating occurs most around the time of nest-building, and the only behavior that has been observed to take place before mating is one in which the female pecks at the male in the neck or shoulder. Following this action the male usually mounts the female and copulation takes place.

Nest-Building

Placement: In buildings or trees, 10–30 feet high
Size: Tree hole or building opening, at least 1½ inches in diameter
Materials: The final nest is lined with dead leaves and grasses

The first stages of nest-building are done by the male during idle moments when there is no immediate need to defend the site. The male first cleans out whatever material is in the nest hole, and then begins to bring in dead leaves, bits of bark, or moss and lichens. In these early stages, he may often be seen gathering material and then just sitting near the nest with it in his bill, and then maybe dropping it. Later in the season, when green leaves or tree flowers are available, the male will bring these into the nest. Often this carrying of material in and out of the nest is done in front of a female, possibly in some way alerting her to the presence of the nest hole and stimulating her to enter it and start building.

When a female has chosen a nest site and a male, she will clean out much of what he has brought into the nest and add her own gathered material, which is mostly grasses. This process takes only a few days, and the male usually perches near the nest hole while this goes on. This final nest-building usually occurs about a week before the first egg is laid. The whole process of nest-building is repeated for successive broods.

Starlings are extremely aggressive birds when it comes to claiming nest sites, and they undoubtedly take over many natural cavities in trees that other birds would use. In some cases birds will be driven out of the nest holes by Starlings and their eggs or young eaten. But this nest predation is not peculiar to Starlings; Blue Jays and House Wrens also do this.

Locating the Nest

WHERE TO LOOK Among older trees at the edges of woods, or in orchards, or around buildings with many nooks and crannies

WHEN TO LOOK Any time of year except in August, when the birds molt

BEHAVIORAL CLUES TO NEST LOCATION Look and listen for males Crowing or Wing-waving, for they almost always do these displays within a few yards of the nest hole.

Breeding

Eggs: 4–5
Incubation: 12 days, by both male and female
Nestling phase: 23 days
Fledgling phase: About 4–8 days
Broods: 1–2

Egg-Laying and Incubation

About a week after the female finishes the nest, egg-laying

starts. The eggs are laid one each day until the clutch is complete. Starling eggs (which are about the same size, shape, and color as Robin eggs) are quite often found just lying on the ground. For some reason they were not laid in the nest, possibly because the female was ready to lay but the nest was not complete or was taken over by another bird.

Both parents take turns with incubation during the day, but at night only the female will remain on the nest. The male flies back to the communal roost each night, returning the next day.

Nestling Phase

Both parents feed the young, brood them, and carry away fecal sacs. The young are brooded for the first week of their nestling life. The female remains in the nest during the night.

Fledgling Phase

How long the young are fed outside of the nest varies, but the period is generally short, for the young soon join other juveniles and form huge flocks that move together about the countryside feeding in open fields and among shrubs. During midsummer these juvenile flocks are a good sign of the breeding stage of your local Starlings. Second broods are common and started around midsummer.

Plumage

Starlings go through one complete molt per year from late July into September. The new feathers are tipped with white or tan, and this gives the birds a speckled appearance in fall and winter. During the fall molt, the birds' bills also change color, becoming a light gray. In spring, just before breeding starts, you will notice two changes in the appearance of the birds. First the lighter tips of their feathers are wearing off, leaving the birds a beautiful, iridescent blue-black. Also their bills become lighter in color, in the case of males, bright yellow.

The young birds that leave the nest this year will be mouse-brown with dark bills until the fall molt, when they will take on the appearance of the parents.

In spring and summer the male can be distinguished by his bright yellow bill with the bluish tint at the base of the lower mandible; the female's bill is lighter yellow and pink at the base of the lower mandible. Throughout the year the male is the only one to do the Crowing and Wing-wave displays.

Seasonal Movement

Starlings are continually extending their range, therefore their migration patterns vary considerably in different areas of the continent. Many Starlings migrate in fall and spring. Generally they join up with other large flocks of similarly gregarious birds such as Grackles, Red-Winged Blackbirds, and Cowbirds.

Throughout the breeding range there are also those that stay through fall and winter in the more northern areas. Certain samples of northern winter populations show that they are often composed mostly of males.

Social Behavior

The habit of forming huge communal roosts in all seasons is a striking habit of Starlings, and one that often comes into sharp conflict with humans. The roosts are usually located in groves of trees or on buildings or bridges. The roost sites are generally fixed for each season. Although the roosts may remain in the same place throughout the year, the birds usually shift location in winter to more sheltered places.

Each morning the birds leave the roost and scatter across the land in small flocks. During the day, usually around noon, these flocks may gather into small "secondary roosts," composed of a number of feeding flocks and located in prominent trees near feeding areas. They disperse in the afternoon to resume feeding. Up to two hours before sunset the birds farthest from the primary roost site may start to fly back. Starlings approach the roost along established flight lines, which are used day after day. Other small flocks join them in midair and the flock size increases as it gets closer to the primary roost site. Some members may drop out and perch at established sites along the way called preroosting sites. These are along flight lines and are constantly changing in membership as birds leave and rejoin the main flocks. Before sunset all birds at preroosting sites will have left for the primary roost, where there are immense flocks swarming above it. The birds often make spectacular dives into the primary roost and then settle down for the night.

Roost sizes vary from a few hundred birds to a few hundred thousand birds, and the noise at a large roost can be heard far away. Roosts are largest in late summer, when they are composed of newly hatched young, their parents, and other birds that did not breed. The roosts become smaller in fall and winter (and may change location), when the adults may be migrating or returning to the breeding grounds. Spring roosts are composed primarily of first-year and unmated males. Breeding males join them when their females are incubating on the nest at night. Then when the nestlings are about twelve days old, the female may also join the roost at night, but return the next day to care for the young.

Feeder Behavior

Starlings are attracted to both seed and suet at your feeders. They are aggressive and keep smaller birds from approaching. Most displays of Starlings are seen in the vicinity of their nest holes; at feeders there will be displays of aggression only.

Most common displays: Bill-wipe, Sidling, and Fluffing. See the Display Guide.

Other behavior: A hole in a nearby tree or building might be claimed as a nest site. Behavior at nests includes Crowing and Bill-waving.

Red-Eyed Vireo / *Vireo olivaceus*

THE MALE RED-EYED VIREO IS EASY TO LOCATE DUE TO HIS CONSTANT singing through most of the breeding period, but he can be hard actually to see, for he tends to remain in the tops of broad-leaved trees where his colors blend with the leaves and dappled light. I always feel a real sense of accomplishment after actually spotting one and keeping him in view for any length of time.

Sometimes he stops singing and comes down from the treetops to interact with his mate. This gives you a rare opportunity to locate the female, for she usually stays in the understory of the forest and does not sing, although she may give one of several calls. Once you have found the female, follow her movements and she will lead you to the nest within a short time, for she does all the nest-building and incubation. Having located the nest, you can follow the female by sight as she comes and goes, and you can keep track of the male through his constant singing. At this point you will be in a perfect position to gain some insights into the workings of these beautiful birds' secretive lives.

Don't miss the experience of watching the female during incubation as she sits absolutely still on the nest while the male continually sings his short phrases from the treetops. At some point he will stop singing, and you will notice an immediate change in the female's behavior. She begins to look in all directions, move about on the nest, and may even give one of her calls. In a minute or two she usually flies off to meet the male and is either fed by him or feeds on her own. After about five minutes the male will be singing from the treetops again and, possibly without your even noticing, the female will have come back to the nest and resumed incubating. From this it will be clear to you that the male's song is being closely listened to by the female and is a continual aural contact that helps the pair coordinate this phase of their lives.

BEHAVIOR CALENDAR

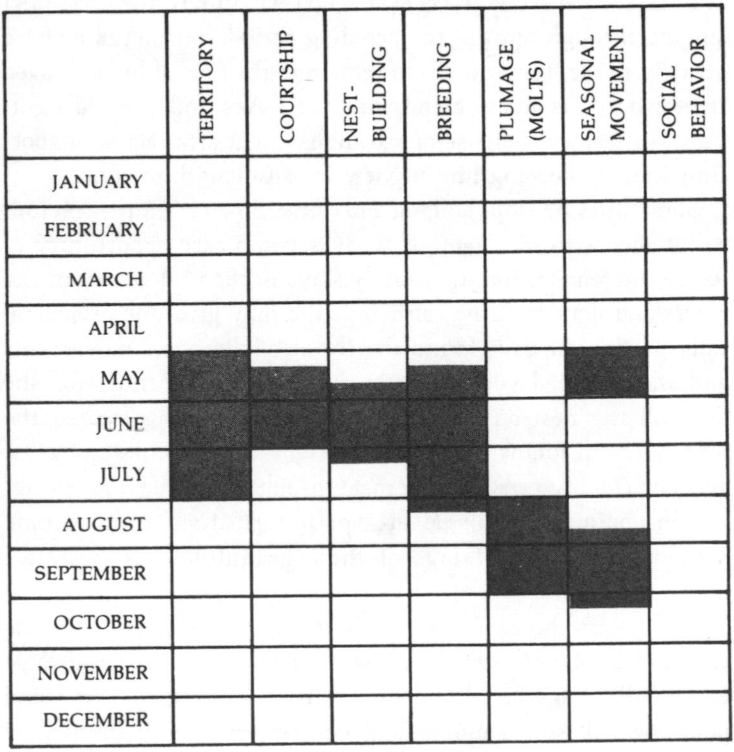

	TERRITORY	COURTSHIP	NEST-BUILDING	BREEDING	PLUMAGE (MOLTS)	SEASONAL MOVEMENT	SOCIAL BEHAVIOR
JANUARY							
FEBRUARY							
MARCH							
APRIL							
MAY							
JUNE							
JULY							
AUGUST							
SEPTEMBER							
OCTOBER							
NOVEMBER							
DECEMBER							

DISPLAY GUIDE

Visual Displays

Crest-Erect

Male or Female *Sp Su*

The bird raises its crest feathers and fluffs its neck feathers, showing a generally alert attitude.

CALL Myaah-call

CONTEXT Occurs in situations of mild aggression or fear; may occur between mates when one suddenly approaches the other. *See* Territory

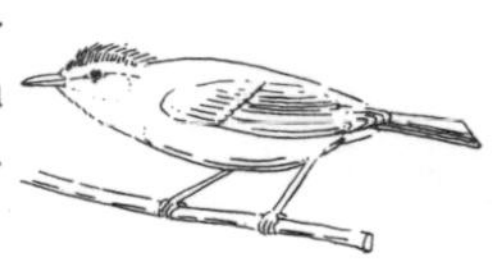

Tail-Fanning

Male or Female *Sp Su*

The tail is fanned and lowered, the body feathers are slightly fluffed, and the crest may be raised.

CALL Myaah-call or Tjjj-call

CONTEXT Occurs in aggressive situations between competing males or between mates. *See* Territory, Courtship

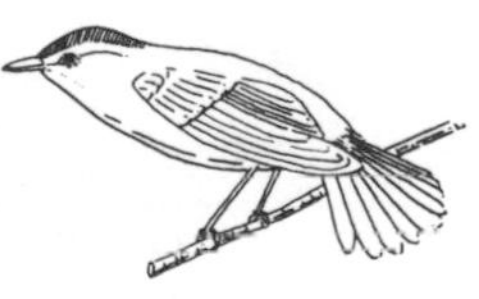

Swaying

Male *Sp Su*

The tail is fanned, body feathers fluffed, and head feathers sleeked. With this posture the bird sways his upper body back and forth.

CALL Warbling-song

CONTEXT Occurs in front of the female, who may act aggressively toward the displaying male. *See* Courtship

Wing-Quiver

Male or Female *Sp Su*

Wings are held out slightly and quivered. In sexual displays by the female, the tail may be raised.

CALL Tchet-call

CONTEXT Done mostly by females either preceding copulation or before and during being fed by the male. *See* Courtship, Breeding

Auditory Displays

Song

Male *Sp Su*

taree. tareo. talio. tarit. etc. Only a sample.

A continual series of two- to three-note phrases, with pauses between phrases; each phrase different from the previous one; given throughout the day

CONTEXT Given by male from the moment he arrives on the territory; about thirty to forty phrases per minute during territory formation; with the start of incubation phrases often given faster — fifty to sixty per minute. *See* Territory, Breeding

Warbling-Song

Male *Sp Su*

The only continuous warbling call of this Vireo; not short phrases with pauses like the Song

CONTEXT Given by the male during close interactions with the female. *See* Courtship, Breeding

Myaah-Call

Male or Female *Sp Su*

A drawn-out, nasal sound *myaahhn.*

CONTEXT Used by male or female with Tail-fanning or Crest-erect; an aggressive call. *See* Territory, Courtship

Tjjj-Call

Male or Female *Sp Su*

The only drawn-out, rasping call of the Vireo; similar to the scolding call of the Northern Oriole *tjjjjj*

CONTEXT Used by male or female with Tail-fanning display or else just by itself in aggressive encounters. *See* Territory, Courtship

Aerrr-Call

Male or Female *Sp Su*

A soft, drawn-out call given between mates; does not sound threatening, as do the Myaah-call or Tjjj-call *aeeerrr*

CONTEXT Given between mates when they meet. *See* Courtship

Tchet-Call

Male or Female *Sp Su*

A harsh, short single sound given repeatedly; not drawn out, as the Tjjj-call is. *tchet, tchet, tchet,*

CONTEXT Given by the adult female when she Wing-quivers and is about to be fed by the male; given by fledglings when begging for food from the adults. *See* Breeding

Mating Calls

Two other calls used only during mating
Male: *tewtewtewtew*
Female: *quotquotquotquot*

BEHAVIOR DESCRIPTIONS

Territory

Type: Mating, nesting, and feeding
Size: 1–2 acres
Main behavior: Song, chases, Tail-fanning, Crest-erect, Myaah-call, Tjjj-call
Duration of defense: From arrival of male to fledgling phase

As soon as the males arrive on the breeding ground, they begin to patrol areas of their own, each about four to five acres in size. Their behavior at this time consists of continual movement about the periphery of the territory, singing their Song phrases loudly but at a slower tempo than they will use later in the breeding season. As more males arrive, the areas patrolled by the earliest birds become smaller and better defined. Conflicts between males over territory generally take the form of short, circular chases alternating with short periods of perch-taking and intermittent song. Encounters may also include the displays of Tail-fanning or the Tjjj-call.

Females arrive a few days after the males and pair with territorial males. At this point territories become more actively defended, and the Myaah-call is more common. The territories now reach their final size of from one to two acres.

Disturbances on the territory generally cause the male to stop singing and approach with Crest-erect. Birds that prey on the nest, such as Crows, Blue Jays, and Common Grackles, are at first followed quietly, but if they approach too near the nest they are openly attacked and called at with the Myaah-call.

Courtship

Main behavior: Chases, Swaying, mating, Tail-fanning, Warbling-song, Aerrr-call, Myaah-call
Duration: Variable

It is hard to follow some of the activities of the Red-Eyed Vireo, because it tends to live and move about in the dense cover of open, broad-leaved forests. You are more likely to see courtship activities in this bird by chance, rather than by patiently sitting in one place and waiting for them to occur. One of the first interactions between mates is a vigorous chase of the female by the male. This may continue for long periods of time and can be hard to follow, since the birds will move in and out of dense cover. Earlier the male was also involved in territorial chases with other males, and in these, both members of the chase could be heard singing. In mate chases the female will not sing, and she persists in remaining on the territory.

There are a number of other behavior patterns believed to be associated with Vireo courtship, and they may be seen alone or in various combinations. One of these is the Aerrr-call, given by either the male or the female when the two meet. This call may be accompanied by Tail-fanning, in which the tail is lowered and spread, and the crest erected and throat feathers fluffed. The Warbling-song may also be given during this display or on its own during male-female encounters.

Swaying is an unusual display that also occurs between mates, and seems to be common to many other species of Vireos. It is done by the male in front of his mate, and may continue over a period of up to fifteen minutes. In it the male fans his tail, ruffles his body feathers, sleeks his head feathers, and sways his body back and forth. The male may give the Warbling-song and the female may be silent or give the Myaah-call.

Mating in Vireos does not differ significantly from that of our other common birds. The female squats down, raises her tail, and Wing-quivers. The male mounts her and copulation takes place. There are special calls of both the male and the female that are given only during this interaction.

Nest-Building

Placement: Usually under 15 feet high, suspended by its rim on forked horizontal branch
Size: Inside diameter 2 inches
Materials: Bark strips, fine grasses, spider egg cases, lined with plant down

The female Vireo selects the nest site by herself while the male sings from elsewhere in the territory. The female does all of the building, while the male just accompanies her to and from the nest as she gathers material. While the female is actually placing material in the nest, the male perches nearby and may give Song, the Warbling-song, the Aerrr-call, or a call associated only with mating. At first the female is hostile to any attempts at mating, but when the nest is nearer completion she is more receptive.

The Vireo's nest is one of the most beautiful to be found. It is made with birch bark strips, spider or caterpillar webbing, filaments from weed stems, paper from wasp nests, and egg cases of spiders. The female gathers the bark strips while perched on the tree or sometimes while hovering beside the trunk. The nest takes four to five days to complete, but all through the incubation phase and even when the young have hatched, she may add material to the outside or rim.

Locating the Nest

WHERE TO LOOK In open deciduous forests, on a horizontal branch 5–15 feet off the ground
WHEN TO LOOK From mid-May through June
BEHAVIORAL CLUES TO NEST LOCATION:

1. Locate the territory.

2. Look for the pair flying about together, for the main time that this occurs is during nest-building.

3. Watch for the female's gathering nest material such as tree bark or silken webbing.

Breeding

Eggs: 3–4
Incubation: 12–14 days, by female only
Nestling phase: 10 days
Fledgling phase: 2–3 weeks
Broods: 1–2

Egg-Laying and Incubation

After the nest is complete, there is a period of two to four days before the first egg is laid. During this time the male and female generally stay near the nest and the female gives the precopulatory displays of Wing-quiver and *quotquot* call. Mating is more frequent at this stage than at any other time.

The eggs are laid one each day until the clutch of three to four eggs is complete. As incubation starts, the behavior of the pair changes. The female does all incubation and the male rarely comes up to the nest, but rather remains in the treetops singing and feeding. His song becomes markedly more persistent than before, with more phrases per minute.

When he stops singing, the female becomes agitated on the nest; she moves about and starts to look in all directions. The male approaches the nest and either he or she may give the Aerrr-call. Then the female suddenly leaves the nest and flies off in the direction of the male. They feed together, the male often feeding the female as she Wing-quivers and gives a call similar to the one the young will give when they beg for food. The female then returns to the nest and the male resumes his singing high in the trees. At times only parts of this pattern may be seen.

Nestling Phase

Once the young hatch, the male changes his behavior and now approaches the nest frequently. He feeds the young directly, or else gives the food to the female, who in turn feeds the young. The female broods the young almost constantly for the first few days after hatching, but by the sixth day all brooding stops.

Early in the nestling phase the fecal sacs of the young tend to be eaten by the adults; later they are carried away and left on twigs or among bark. Near the end of the nestling phase, the male spends more time in the singing area and the female feeds the young herself.

Fledgling Phase

Once the young leave the nest, territorial boundaries are no longer defended, and the parents with their young may roam widely. There is evidence that the young may be fed by the parents for up to two weeks after fledging, but the families can be hard to locate.

Plumage

Red-eyed Vireos go through one complete molt per year, in August and September, just before the birds migrate south. During the molting period the male stops singing.

There are no visible differences between male and female coloration in the Vireo; therefore any attempt to distinguish the sexes must use behavior. The main clues are that the male is the only one to give Song and that the female does all the incubation.

Seasonal Movement

Male Red-Eyed Vireos migrate north in spring a few days before the females. They arrive in the middle latitudes in late April and early May, and immediately begin their Song. After the molt in late summer, the males may resume singing for a short time and then they and the females start migration south.

FL

Common Yellowthroat / *Geothlypis trichas*

COMMON YELLOWTHROATS GENERALLY BREED IN MOIST, SHRUBBY areas, and because of their small size and protective coloration they can be very hard to observe. Luckily for behavior-watchers, the male frequently sings, and his song is both loud and easily recognized. The male arrives on the breeding ground about a week before the female and immediately starts to establish a territory. The female's arrival may go unnoticed, for she is inconspicuous and often remains out of sight. But by watching the male you can guess when it occurs, for as soon as she arrives, he stops almost all singing and follows her closely as she moves about the territory.

During this period the female builds the nest. It is built close to the ground and composed of three layers of successively finer grasses. The nest is difficult to locate, even though the female has a tendency to give a scolding call if you get too near it. Once the nest is completed, the male stops following the female and resumes giving his Song.

During the breeding period you may see the male do an unusual display. He flies twenty-five to one hundred feet up with shallow looping flight, giving sharp calls as he ascends. At the top of the flight he gives a harsh warbling call that may have some similarities to his Song, and then descends quietly. The function of this display is not known, but it does occur over the territory, and the female has been observed to take cover whenever it is performed.

After the eggs are laid, the female incubates them for about twelve days. The nestling phase is surprisingly short, lasting only eight days before the young move out of the nest and perch in cover nearby. Soon they can fly well and follow one or both parents about the territory as they continue to be fed by them. The female may leave the fledglings in the care of the male as she starts a second brood.

The male stops territorial singing around the beginning of August, for this is when Common Yellowthroats molt. Following this the males leave first on their southern migration, followed in a week or two by the young and the adult females.

BEHAVIOR CALENDAR

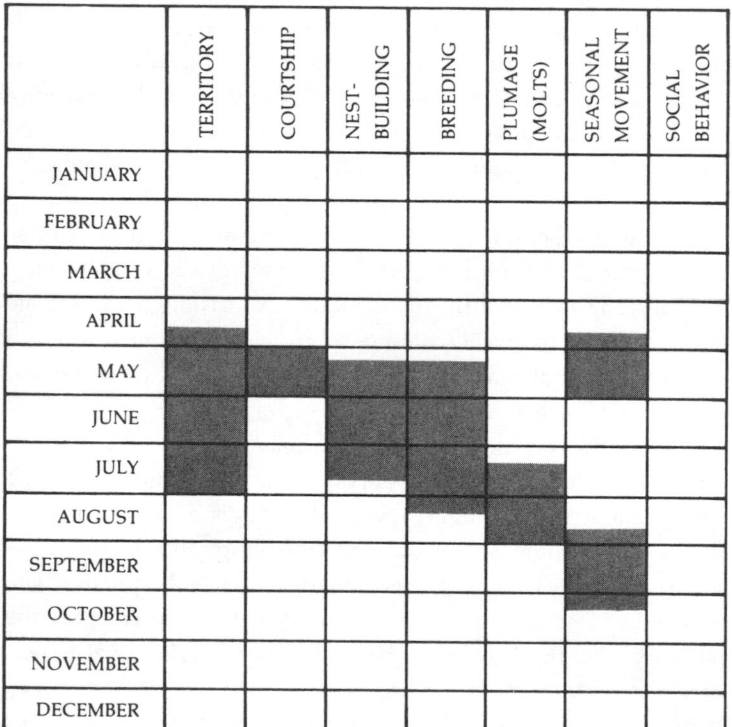

	TERRITORY	COURTSHIP	NEST-BUILDING	BREEDING	PLUMAGE (MOLTS)	SEASONAL MOVEMENT	SOCIAL BEHAVIOR
JANUARY							
FEBRUARY							
MARCH							
APRIL							
MAY							
JUNE							
JULY							
AUGUST							
SEPTEMBER							
OCTOBER							
NOVEMBER							
DECEMBER							

DISPLAY GUIDE

Visual Displays

Song-Flight

Male *Sp Su*

The bird flies up into the air with looping flight while giving a few sharp calls like *teenk, teenk, teenk.* At the height of his flight, twenty-five to one hundred feet, he gives a variable warbling call and then descends quietly to a perch.

CALL Song-flight-call

CONTEXT Done by a male over the territory; most common in the afternoon and late in the breeding season; tendency of female to disappear when the display occurs. *See* Territory

Wing-and-Tail-Flick

Male *Sp Su*

The bird moves about with nervous darting flights, flicking his tail and wings when briefly perched between flights.

CONTEXT Done by neighboring males when first forming territory and is preceded by the scolding call; also occurs during courtship period when male is attempting to induce the female to copulate. *See* Territory

Auditory Displays

Song

Male *Sp Su*

wititee wititee wititee A three- or sometimes two-syllable phrase continually repeated; variable within the species but distinct in each individual

CONTEXT Given by male during most of the summer from various perches in his territory, in order to advertise the territory to females in early spring and to males throughout the breeding season. *See* Territory, Courtship

Stetedeet-Call

Male or Female *Sp Su*

stetetedeet or *stedeet* A harsh, rapid call, usually drawn out but sometimes shortened. Tail of bird often vibrates when call is given.

CONTEXT Directed at a bird's mate, either to call it or to warn it; also used during vocal duels between males in spring when territorial boundaries are being disputed. *See* Territory, Nest-building

Tchat-Call

Male or Female *Sp Su*

tchat. tchat. A short, single sound sometimes repeated

CONTEXT Given by male or female when anything out of the ordinary occurs within their territory

Song-Flight-Call

Male *Sp Su*

Two-part call: the first part made up of short harsh notes; the second, a variable warble

CONTEXT Accompanies the Song-flight visual display: the harsh notes during the flight ascent, and the warble at the peak of the flight — *teenk, teenk, teenk,* followed by a variable warble

Steek-Call

Male or Female — *Su*

A sharp, repeated call

CONTEXT Given when an observer is close to the nest; a higher intensity than that of the Tchat-call. *See* Nest-building — *steeek. steeek.*

Zeeyeet-Call

Male or Female — *Su*

A soft call much like its written description — *zeeyeet. zeeyeet.*

CONTEXT Given by the birds as they approach the nest; occurs during the nestling phase

BEHAVIOR DESCRIPTIONS

Territory

Type: Mating, nesting, feeding
Size: Approximately 1–2 acres
Main behavior: Singing from exposed perches, vocal duels, chases
Duration of defense: From arrival on breeding ground until nestling phase; starts again for second brood

Male Yellowthroats start territory formation almost as soon as they arrive on their breeding grounds. Territory formation is obvious, taking the classic form of the male's singing frequently and loudly from various song perches within his territory. He watches the activities of other nearby males carefully, and when one intrudes or comes too close to a border, the two males meet and do short flights around each other and give the

Wing-flick display. This activity may be preceded by the Stete-deet-call, which seems to function as a warning. In contrast, the short flights and Wing-flick display are done in silence.

Males that arrive first rarely have difficulty in establishing territories. Late arrivals may crowd the original birds as they define their own territories, thus creating more conflict between neighboring males.

The size of Yellowthroat territories seems to vary widely, and there is some question as to whether the birds also occupy a range, only part of which is defended as a territory. Sizes of territories generally average from one-half to two acres. One confusion may arise from the fact that from the nestling phase on, the adults do not defend their territories and move freely in search of food. Before the second brood starts, territory is again established and defended.

A fairly common display, especially later in the breeding season, is Song-flight. In this display the male flies up at an angle into the air for twenty-five to one hundred feet. During the ascent, sharp call notes are given. At the end of the upward flight the bird gives off a harsh, warbling call and then descends silently. The function of the display is not clear; it may be an added territorial display, or some form of distraction display in the presence of possible predators.

Courtship

Main behavior: Absence of male song; the pair flying about together; female wing-quivering
Duration: About 1 week

Pair formation in the Common Yellowthroat is not conspicuous and seems to occur with the aid of very few displays, but there are a few behavioral clues that indicate when pair formation is taking place.

The first is that the male, who has been singing constantly during the previous weeks of territory formation, suddenly stops all Song. Some observers have found this so closely associated with the arrival of the female on the territory that they use it to date her arrival, for the female is extremely secretive and well camouflaged, and her presence cannot always be discerned by sight.

A second sign of pair formation is that the two birds will be seen moving about together on their territory. This would not be the case if the bird were not the mate of the male, for another male would never be tolerated within the territorial borders.

During this time the female may also wing-quiver, which, as in many other passerines, seems to signal her readiness for copulation. Her visual display is accompanied by a high, peeping call. Some observers have found the female to accept any male that approaches her as she displays and not just her mate. Sometimes her displaying may attract other males, and her mate may have to chase them off.

During this period of courtship, the female is also looking for a nest site, and once she begins building, the male spends more time away from her and resumes vigorous singing.

Nest-Building

Placement: In brush areas of open fields, from ground level to 2 feet high
Size: Inside diameter 1¼ inches
Materials: Made of coarse grasses and dead leaves, lined with fine grasses or hair

The search for a nest site and the building of the nest both take place within the courtship period. The female does all the building and tends to be extremely secretive as she does so. The male may accompany her on her trips to gather nesting material, but as they return, the male approaches no closer than about twenty feet from the nest. Therefore, the male will rarely lead you to the nest site during building. The nest takes two to five days to complete, but following this there may be a period of up to a week before eggs are laid.

Second nests or nests for second broods tend to be slightly higher, possibly because of the increased summer vegetation near ground level.

Locating the Nest

WHERE TO LOOK In areas where there is a mixture of low shrubbery and higher growth; often near water: lakes, streams, etc.
WHEN TO LOOK May through July
BEHAVIORAL CLUES TO NEST LOCATION:

1. Look for the nest during the courtship period.
2. A female will declare herself with a scolding call if you get too near the nest.
3. When you approach the territory, the male may dive at the female, and she will then leave the vicinity of the nest and not continue building until you are out of sight.

Breeding

Eggs: 3–4
Incubation: 12 days
Nestling phase: 8–9 days
Fledgling phase: 2–3 weeks
Broods: 1–2

Egg-Laying and Incubation

After completion of the nest, the female may wait up to a week before laying the first egg. After starting, she lays one egg each day until the clutch is complete.

The female's periods of incubation last forty-five to seventy minutes at a time, and then she leaves the nest to feed. The male may accompany her to and from the nest, but he does not approach the nest any closer than about three feet at this stage. When the female approaches the nest, she first lands a short distance from it and then continues by hopping through the underbrush.

Common Yellowthroats are frequently parasitized by Cowbirds. The Cowbird lays an egg in the Yellowthroat nest, it hatches, and the young Cowbird is raised by the adult Yellowthroats.

Nestling Phase

Both parents feed the young, but sometimes the male will bring the food to the female, who will then pass it on to the young. The young spend a relatively short period in the nest, leaving it, before they can fly, by hopping and walking into nearby shrubbery.

Fledgling Phase

Because the young leave the nest and are separated before they can fly, they must remain under cover while the parents feed them. After three days like this, they are able to fly and will all join together in following the parents about. The family group may move up to one thousand feet from the nest site. About twelve days after leaving the nest, the young can fly well and spend more time feeding on their own. About twenty days after leaving the nest they no longer remain near the parents. The female parent may leave the family group early to start a nest for the second brood.

Second broods may occur in southern areas; a new nest is built and territorial boundaries may be slightly changed.

Plumage

Common Yellowthroats go through one complete molt per year, usually in July and August. Their coloring is the same throughout the year.

The sexes are easily distinguished by plumage, the male being the only one to have a black band across his eyes and forehead.

The male is also conspicuous, for he often sings and frequently sits on exposed perches. The female is usually very secretive.

Seasonal Movement

All the males arrive on the breeding ground within a week or less of each other. The females arrive seven to ten days later, but may not be seen, since they are more secretive in their habits and are not involved with territorial advertisement.

In the fall the males leave a week or two before the females. Migration north occurs in April and May. Migration south occurs from late August into October.

House Sparrow / *Passer domesticus*

YOU WILL FIND THE HOUSE SPARROW AN ENGAGING SUBJECT FOR behavior-watching once you start to take an interest in distinguishing its various calls and watching its interactions with its own and other species.

One common and conspicuous behavior of a group of House Sparrows is seen when the birds rush noisily in among some plants and hop excitedly about. If you look closely, you will see that these groups consist of one female and many males, and that the males are continually hopping and bowing and sometimes pecking at the female. The rapid chirping calls that accompany this behavior are often the best clue to its occurrence. The display is believed to be associated with courtship and is described more fully in the behavior descriptions.

As you walk through the city, you may see a male House Sparrow perched on a store sign or air conditioner, chirping repeatedly. This is the typical behavior of a male with a nest site who is attempting to attract a mate. The display indicates not only the breeding stage of the male, but also the fact that there is a possible nest site within a few feet. When a female arrives, the male may excitedly fly to the nest site, possibly stimulating the female to follow.

The roosting of House Sparrows is another prominent aspect of their behavior, and undoubtedly you know some ivy-covered wall or group of dense plantings where House Sparrows gather in great numbers during fall and winter evenings. The birds also gather regularly at fixed spots during midday, chirping noisily for up to an hour. These are secondary roosts, and, like the primary roosts, they are used mainly in fall and winter.

You can see practically all stages of the breeding cycle anytime during the spring or summer, for each mated House Sparrow has a number of broods. Although breeding takes place in spring and summer, you may see some examples of preliminary nest-building and courtship through the fall and winter.

BEHAVIOR CALENDAR

	TERRITORY	COURTSHIP	NEST-BUILDING	BREEDING	PLUMAGE (MOLTS)	SEASONAL MOVEMENT	SOCIAL BEHAVIOR
JANUARY							
FEBRUARY							
MARCH							
APRIL							
MAY							
JUNE							
JULY							
AUGUST							
SEPTEMBER							
OCTOBER							
NOVEMBER							
DECEMBER							

DISPLAY GUIDE

Visual Displays

Hop-and-Bow

Male *Sp S F*

One or more males hop about in front of a female; heads held high, tails up and fanned, wings drooped slightly, and gray rump feathers fluffed. In between hops they bow stiffly.

CALL Chirp-call

CONTEXT Occurs in fall and in spring, usually started by a mated pair; female aggressive to the males and as she flies away, they follow; ends as suddenly as it starts. *See* Courtship

Wing-Quiver

Male or Female *Sp Su*

A bird crouches, opens its wings slightly and quivers them rapidly.

CALL Nest-site-call or others by female

CONTEXT Done by male during courtship when a female first approaches; done by female when initiating mating; done by male outside the nest just before the nestlings fledge; done by fledglings as they beg for food. *See* Courtship, Breeding

Head-Forward

Male or Female *Sp Su F W*

The bird places its body horizontally with head forward and wings slightly spread. Its bill may be open.

CONTEXT Given as a close-distance threat to another House Sparrow or other bird; preliminary to a fight if it does not make the other bird move away; often seen at feeders. *See* Territory

Tail-Flick

Male or Female *Sp Su F W*

The bird sits erect and repeatedly flicks its tail.

CALL Quer-quer-call

CONTEXT Given at times of possible danger, as when a strange House Sparrow or other bird is near the nest. *See* Territory

Auditory Displays

Nest-Site-Call

Male or Female *Sp Su*

chirup chireep, chirup.

A two- to three-part phrase that is rhythmically repeated; sounds as if all parts of the phrase start with "ch" and end with "p."

CONTEXT Given by the male when he is perched near the nest site and has not yet attracted a mate, or by a female that has lost her mate. *See* Courtship

Chirp-Call

Male or Female *Sp Su F W*

cheeup or *chirup* or *cheeleep*

The typical, single-note call of the House Sparrow; extremely variable

CONTEXT Given by the male during Hop-and-bow; given by male and female during night or day roosts. *See* Breeding

Churr-Call

Male or Female *Sp Su*

A call continuing for one to two seconds, which has the quality of a rapidly repeated "ch" sound

CONTEXT Given mostly by the male as he chases an intruder or predator away. *See* Territory

Quer-Quer-Call

Male or Female *Sp Su F W*

A nasal two-part phrase given twice in quick succession and then repeated several times; loud and persistent

querquer. querquer. or *kewkew, kewkew,*

CONTEXT Given at times of possible danger. *See* Territory

BEHAVIOR DESCRIPTIONS

Territory

Type: Nesting
Size: Immediate area of nest site
Main behavior: Churr-call, Head-forward
Duration of defense: Variable, in some cases throughout the year

Territorial behavior in the House Sparrow is basically defense of the nest site. There is no defined area outside the nest that the bird consistently defends. Interactions concerning ownership of nests occur mostly in spring and late fall, the main times when House Sparrows claim nest sites. In skirmishes around the nest, generally males chase males and females chase females. The main displays used are the Churr-call and the Head-forward. Both function as threats toward other birds, and if these fail in their purpose, actual chasing may occur.

There are many other bird species that compete with House Sparrows for nest sites. In the city there are Starlings; in more rural areas there are House Wrens, Tree Swallows, and Bluebirds. When the House Sparrow is interacting with one of these species it will use the Churr-call, Head-forward, and also the Quer-quer-call. These same calls are used during breeding when predators are in the area of the nest.

Courtship

Main behavior: Nest-site-call, Hop-and-bow, Wing-quiver
Duration: Variable, seen intermittently throughout the year

One of the first stages of House Sparrow courtship is the Nest-site-call, given by the male right near his nest site. This is often seen in the city — the male perched on a roof gutter, a store awning, or an air conditioner. The constant rhythmic quality of the call is the best way to recognize it. When a female flies near, the male calls faster and louder and starts to Wing-quiver. If the female starts to fly away, the male may follow after her for a short distance, continuing the Wing-quiver and the Nest-site-call. If the female stays nearby, then the male may start to go in and out of his nest. After this the female may also enter the nest, even though the male may be slightly aggressive at first. A paired male may continue to give the Nest-site-call but in a softer manner and never accompanied by the Wing-quiver.

This behavior is common in spring but also occurs to some extent in fall. A male that is unable to attract a female (possibly because of an unsuitable nest site) may keep doing the Nest-site-call incessantly through much of the summer; the call often gets louder and harsher the longer he goes unmated. Females or males that lose their mates in the middle of the breeding season also will start the Nest-site-call to attract a new mate, for the birds are generally faithful to their nest sites throughout their lives. A female that loses her mate and does not soon gain a new one through display may leave her nest site to join a calling male.

Another common courtship display is the Hop-and-bow. Most people have seen this display, even though they may not have recognized it. What we often see when the display takes place is a group of noisy sparrows dashing after each other and landing in a dense hedge or bush where they give rapid *chirp* calls and hop around; then the whole group may pick up and dash to another spot where the same frenzied activity takes place. The rapid *chirp* calls are distinctive and can lead you to the display. The display group consists of one female and the rest males. The female leads the group as it moves from spot to spot, and the males chase after her. When the birds land, the males Hop-and-bow as they move about in front of the female. They may also peck at her. The female responds by standing erect with bill open, and by lunging toward the males.

The display may involve just a single female and her mate, but if the female flies and the mate follows her, calling loudly, this is believed to attract other males, resulting in a large number of birds' participating in the display. The Hop-and-bow, along with the chasing, can be observed in almost any month of the year.

Mating is commonly seen in House Sparrows. The female always initiates it by crouching down, drawing her head into her shoulders, and quivering her wingtips. The accompanying call is a high-pitched *teeteeteeteetee*. The male responds by hopping toward the female. He lands on her back and bends his tail down to make contact with her. He then gets off and either bill-wipes or preens. If she continues to Wing-quiver, he hops on her back again. The birds may touch bills while copulating, or the male may peck at the neck feathers of the female. Mating occurs throughout the breeding cycle, usually near the nest site, and may occur many times during the day.

Nest-Building

Placement: In the crannies of buildings or in trees, 10–30 feet high
Size: Spherical and 8–10 inches in outside diameter
Materials: Grasses, leaves, twigs, cloth; lined with feathers or fine grasses

The nest plays a very important role in the lives of House Sparrows, for the birds use it as a center of activity for almost the entire year. From spring through summer it is used for raising young, sometimes up to four broods, and then in fall through winter it is used as a resting place during the day and as a roosting site at night.

House Sparrows most often nest in the nooks and crannies of human habitations, but when there is a lack of these sites they may also build in trees. The exposed nests are spherical in shape and about the size of a soccer ball. Groups of nests in trees are more common in Europe than in this country, but one North American researcher reported seventeen nests in a single tree.

House Sparrows claim nest sites at two different times of the year. The main one is in spring just before breeding, but there is another period in fall after the birds have left their late summer communal flocks and are returning to the breeding area for the winter. At this latter time the adults return first and reclaim their past sites, then the juveniles return and compete over the remaining sites.

Since the nests are used all year, except for a brief period in late summer, you can always see some nest-building activity — either repair, cleaning out of old nest material, or building of new nests. The main times of building are in spring and briefly in fall. Both male and female participate in the building, although an unmated male may begin a nest before advertising for a female. The birds can be quite amusing to watch as they gather material, for they often pick at bits of objects that are hard to break loose, and may spend up to ten minutes working on a single twig.

In winter, one or both members of a pair may spend the night in the nest for protection from the cold.

Locating the Nest

WHERE TO LOOK In the nooks and crannies of buildings
WHEN TO LOOK In spring, summer, or late fall
BEHAVIORAL CLUES TO NEST LOCATION:

1. Look for a male giving the Nest-site-call.
2. Follow Sparrows carrying nesting material.
3. Listen for the Churr-call, for it is usually given when there is danger near the nest.

Breeding

Eggs: 4
Incubation: 12 days, primarily by the female
Nestling phase: 15–17 days
Fledgling phase: About a week
Broods: 2–3

Egg-Laying and Incubation

Egg-laying starts about a week after active nest-building occurs. The eggs are laid one each day in the early morning. During the days that the eggs are laid, more nest material may be brought in, especially the soft lining of feathers or fine grasses. This lining may help regulate the temperature of the eggs before real incubation begins.

The female starts incubation around the time the next-to-last egg is laid. Only the female develops a brood patch, so she is probably the only one actually to incubate the eggs. The male, however, will stay with the eggs during the times when the female leaves the nest to feed. The female stays with the eggs at night.

There are a number of calls given between the pair whenever they meet at the nest or when they exchange places in the nest:

one is a soft *churrrr*; another is like the words *tew tew* or *chew chew*.

Nestling Phase

Generally the young all hatch on the same day. They are fed by both the male and the female, and in many cases a third bird may join the pair in feeding the young. (In one study, this occurred in sixty percent of the nests.) Two to three days before the young leave the nest, their behavior radically changes. Instead of being noisy and active, they are quiet and stay crouched at the back of the nest except when being fed. On the day of fledging, the adults may not feed the young until the fledglings leave the nest.

Near the end of the nestling phase, the male stops feeding the young and perches near the nest, giving an almost continuous version of the Chirp-call that sounds like *chileep*. When the female arrives with food for the young, he calls louder and even Wing-quivers slightly. When the young leave the nest the male stops this behavior.

Nestling House Sparrows are often preyed upon by other birds, especially Blue Jays and Crows. In these situations the parents give loud versions of the Churr-call and the Quer-quer-call.

Fledgling Phase

The young generally leave the nest all on the same day. They will fly after the parents and "beg" for food with the Wing-quiver and a high, repeated call. They may beg from any adult, but generally only their own parents will feed them. In some cases the female may start renesting immediately, in which case the male does all feeding of the fledglings.

Once the young can feed on their own, they join roaming flocks of juveniles that feed and roost together.

Plumage

House Sparrows go through one complete molt per year in late summer. The new feathers are tipped with a lighter brown, but the tips tend to wear off through the winter. This does not affect the female's plumage much, for she is already all brown, but for the male, the lighter tips of the new feathers cover his black bib in winter. By spring the tips have worn off and his bib is large and deep black. Even though the bib of the male is partially covered in fall and winter, it is still the best way to distinguish male from female throughout the year.

Seasonal Movement

House Sparrows are in general not migratory in North America; however, they do have some marked seasonal movements. The main movement occurs at the end of summer when breeding is finished and molting is about to begin. At this time adults and young often leave the breeding area and form large communal flocks that feed and roost together.

The birds return to their breeding areas at two different times. In the colder regions, House Sparrows return to the breeding grounds in late fall and often claim nest sites that they will use for roosts throughout the winter. Northern birds without nest sites roost as small flocks in protected areas. House Sparrows in warmer climates continue their pattern of communal flocking and feeding, which was started in late summer, and return to the breeding ground the following spring.

Social Behavior

House Sparrows are extremely social birds throughout all aspects of their lives. But perhaps one of their most intriguing social patterns is their habit of forming large communal roosts at night and smaller roosts during the day.

In midsummer, when the first young have fledged and become

independent, they form flocks that feed and roost together. Later in the summer these flocks are joined by more juveniles that have fledged, and finally by the adults after they have finished all of their broods.

The flock's behavior is similar to that of many other birds that have communal roosts. The birds may travel from a distance of up to four miles to join the roost at night. Before landing in the roost, they often gather at a preroosting site. These sites are conspicuous, and the birds are noisy and restless at them. From the preroosting site, groups of birds continually break away and go to the primary roost, where again there is much noise and shifting. House Sparrows settle in their primary roosts at night about a half hour before sunset, but may begin gathering at preroosting sites up to an hour and a half before that. The birds leave the roost soon after sunrise.

Other roosts are established during the day, and especially around midday. These are secondary roosts, where the birds sit, preen, and call noisily and continuously. These roosts are used only for an hour or two, are generally small (one hundred birds or so), and are in bushes near the feeding area.

One theory about these roosts is that they are information centers that help birds locate food. In each roost there are birds that have found food and those that haven't. It is believed that birds that haven't found food follow those that have to their feeding areas the next day. The larger the roosts, the greater the area surveyed for food, and the better the chances of survival for the whole flock. The preroosting sites may help advertise the location of the roost and may also be part of the information transfer that helps birds distinguish which ones have found food and which haven't. *See* Starlings and Crows: Social Behavior.

Feeder Behavior

House Sparrows are attracted to the smaller seeds at feeders. They are aggressive birds, and while feeding may keep some other species away. They will be the main visitors at any city feeder. Look for signs of their courtship from the first of the year on into late summer. It takes the form of a noisy flock of birds all dashing to a shrub or tree branch, where they give the Chirp-call and do Hop-and-bow. You may also see the birds form a midday roost, where they remain perched in one spot and give the Chirp-call. See the Display Guide and the sections on Courtship and Social Behavior.

Most common displays: The Head-forward display is used in aggressive encounters at the feeder; the Chirp-call and Hop-and-bow are used away from the feeder in other behavior. Scc the Display Guide.

Other behavior: If House Sparrows come to your feeder, there is a good chance that they will also nest somewhere nearby, such as in a tree hole or on a projection from a building. If they do, you will be able to watch all of their breeding behavior. See the sections on Territory, Courtship, and Nest-building.

FL

Red-Winged Blackbird / *Agelaius phoeniceus*

THE MOST SUCCESSFUL WAY TO OBSERVE THE BEHAVIOR OF RED-winged Blackbirds is to locate a marshy area where a number of them can be regularly found, pick one or two of the more active birds, and follow their movements for about a half hour. A characteristic of Redwings is that they alternate periods of active displaying with periods of quiet and feeding, so you cannot just show up at a marsh and expect immediately to see all of their marvelous displays.

The most exciting behavior of Redwings centers on their territorial defense and the relationship between mates. Males are territorial as soon as they arrive on their breeding ground. Start watching at this time, especially in the morning. If you focus on one or two males, you will learn the extent of their territories and will see how they use their basic displays of Songspread, Bill-tilt, and Song-flight.

A week or two after the females arrive is the most active time on the marsh, and again it is best to take a while to observe one or two especially active territories and try to follow what is happening. It is common to see a female chased by her mate. At times males from neighboring territories may join in the chase and then return to their own territories when it is over. Male Redwings are generally polygamous, so you may see more than one female in a territory.

Later in the breeding season your presence near the nesting area will elicit constant warning calls from adults, and you will have to move quite far away before the birds will go about their normal activities.

Sometime in late July or early August you will suddenly realize that the Red-Winged Blackbirds are no longer on their breeding ground. In fact, the chances are that you will not see a Redwing for the last month of the summer, for this is the time when they join other Redwings in secluded marshes and go through their molt. In September you will see them again, for their molt is finished, and they now feed and roost together in large flocks. In a few weeks they start their migration south.

BEHAVIOR CALENDAR

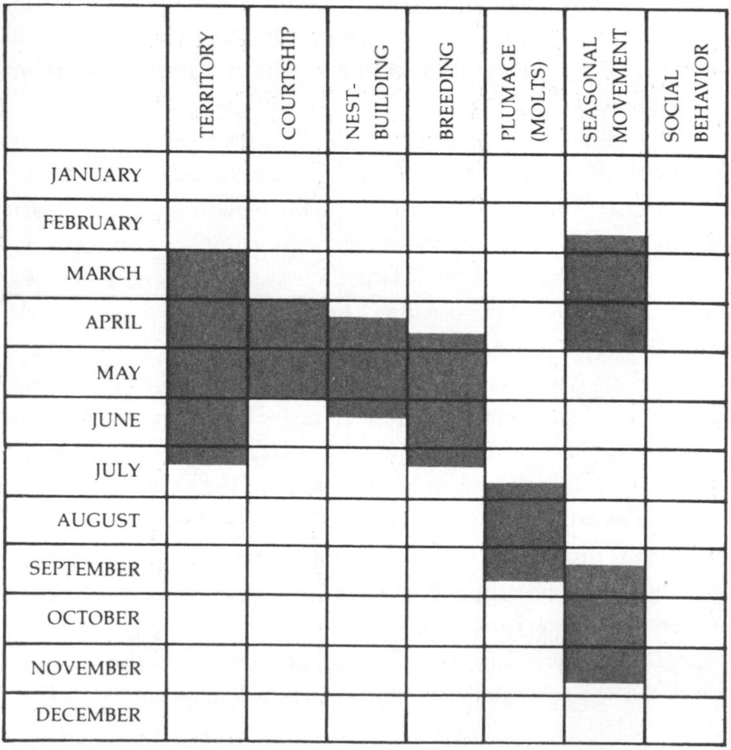

	TERRITORY	COURTSHIP	NEST-BUILDING	BREEDING	PLUMAGE (MOLTS)	SEASONAL MOVEMENT	SOCIAL BEHAVIOR
JANUARY							
FEBRUARY							
MARCH							
APRIL							
MAY							
JUNE							
JULY							
AUGUST							
SEPTEMBER							
OCTOBER							
NOVEMBER							
DECEMBER							

DISPLAY GUIDE

Visual Displays

Songspread

Male or Female *Sp Su*

The bird arches forward, spreads its wings to the side, and exposes the red epaulets. Its tail may be bent down and spread. Display in the female is more subtle, for she lacks the brightly colored epaulets.

CALL Song by male

CONTEXT Given by males during territory formation and when the females first arrive on the breeding ground; given by females when defining their own subterritories. *See* Territory, Courtship

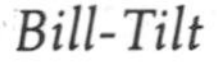

Bill-Tilt

Male or Female *Sp Su*

Two birds near each other both lift their bills above the horizontal plane. The epaulets are exposed in the male version. After the display, one bird generally flies off while the other remains.

CONTEXT Occurs usually between two males at the boundary of a territory. *See* Territory

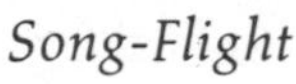

Song-Flight

Male *Sp Su*

The bird's slow, stalling flight, with epaulets exposed, tail spread, and head downward; sometimes ends with a glide to a perch.

CALL Song or none

CONTEXT Occurs when male glides from one perch to another within his territory; affirms the male's presence on the territory. *See* Territory, Courtship

Tail-Flick

Male or Female *Sp Su*

While perched, the bird flicks its tail sharply upward. The tail is spread slightly while flicked.

CALL Check-call

CONTEXT Given especially when possible predators intrude upon the territory; a common response to human presence. *See* Breeding

Crouch

Male *Sp*

The bird arches over, spreads its wings slightly, and holds that position for about five seconds. The Crouch is similar to a low-intensity Songspread but is not accompanied by Song.

CALL None or Teetee-call

CONTEXT Given by the male from a perch above the female in the early stages of courtship. *See* Courtship

Sexual-Chase

Male(s) and Female *Sp*

One or more males rapidly chase a female, who is flying with darting, elusive flight.

CALL Chase-call

CONTEXT A stage of courtship that seems to bond a male and his mate even though other males may join them; sometimes crosses territorial boundaries. *See* Courtship

Wing-Flipping

Female *Su*

While perched, the bird quickly raises and lowers one of her wings.

CALL Check-call

CONTEXT Occurs especially when the female is returning to the nest with food and there is some danger nearby. She then does not go to the nest, but waits and Wing-flips. *See* Breeding

Auditory Displays

Song

Male *Sp Su F W*

A variable phrase except for the last part, which always gets the emphasis and is a drawn-out *eeee*; sounds much like its written description.

ookalee.

CONTEXT Given by males all year; accompanies Songspread and Song-flight during the breeding season. *See* Territory, Courtship

Ch'ch'chee-Chee-Call

Female *Sp Su*

Variable, but characterized by a series of short, harsh sounds followed by milder elongated sounds

ch'ch'ch'cheecheechee

CONTEXT Most common call of female; given in a variety of situations

Check-Call

Male or Female *Sp Su F W*

check or *chuck* — A single harsh sound repeated irregularly; sometimes has a pitch, sometimes is without pitch

CONTEXT Used in connection with Tail-flick and in situations where there is danger on or near the territory. The call of one bird often makes others approach and also call. *See* Breeding

Tseert-Call

Male *Sp Su*

tseeert or *tseeer* or *tseeyeet* — A clear, short whistle often descending slightly; variable

CONTEXT Given when predators are in the area, often in response to aerial predators, in which case all singing and other calls stop momentarily. *See* Breeding

Chase-Call

Male *Sp Su*

tch tch tch tch tch — A rapid series of short harsh sounds

CONTEXT Given by males during Sexual-chase and aggressive interactions with other males. *See* Courtship

Teetee-Call

Male or Female *Sp Su*

teeteeteeteetee — A rapid series of high, short whistles sounding like whimpering, sometimes descending in pitch

CONTEXT Heard from either male or female during courtship or mating. *See* Courtship

BEHAVIOR DESCRIPTIONS

Territory

Type: Mating, nesting
Size: 1/8–1/4 acre
Main behavior: Songspread, Bill-tilt, Song-flight, Tail-flick
Duration of defense: Spring through early summer

Male Redwings arrive on the breeding ground several weeks before the females, and during these early weeks they establish territories. Three visual displays are used in territorial activity.

The main one is Songspread, given from prominent perches within the territory and usually directed at other males, in the period before the females arrive. The display is accompanied by the familiar *ookalee* Song and can be given at varying intensities as indicated by how much is visible of the red epaulets. When the display is at its most intense, the epaulets are fully exposed and are even vibrated.

As males leave or reenter their territories, they often do a second territorial display: Song-flight. A male also does this as he flies between perches in his territory. It is a slow, stalling flight in which the epaulets are flared, and it may be accompanied by Song.

Where neighboring males have a common border, they may meet at the site and do the Songspread, but more commonly do a third territorial display: the Bill-tilt. This display also includes exposed epaulets, but takes place without sound.

Knowing these three displays will help you to estimate the extent of a male's territory, especially during an active period when many males are displaying at the same time. There are two things that make Redwing territory slightly less apparent: the males often drop down among the reeds, out of sight; and they may leave their territory to feed elsewhere a couple of times a day. Generally the most active times on the territory are morning and late afternoon.

Since Redwing territories are usually small and adjacent, there is often a great deal of jockeying for sites, and this can be followed over the weeks but not instantly; you may need to spend a half hour or so to get an idea of which males are claiming the different areas of the marsh.

When the females arrive, they form subterritories within the male's territory. Defense of these areas may involve chases, calls, Songspread and Bill-tilt; but since the females are often busy with nesting, they interact less than the males.

Courtship

Main behavior: Song-flight, Songspread, Crouch, Sexual-chase
Duration: Over a period of three weeks with each mate (males are polygamous)

Females arrive on the breeding ground in small groups a few weeks after the males. They often perch in trees around the periphery of the territories; and males, immediately aware of their presence, may fly up to them, do intense versions of Songspread, and then glide down to their territories with Song-flight. The females may fly on, or they may enter territories after a while.

Three types of interactions commonly occur between the male and the female, from the time she enters the territory up until egg-laying. The male may do the Crouch display as he is perched near the female. The female is often down among the reeds, and the male's display from a perch above her can alert you to her whereabouts. From this posture he may extend his wings and give the Teetee-call. If you have been listening during the period before the arrival of the females, this call is a conspicuous addition to the auditory displays and is a good indication that pair formation has started.

A second interaction involves an aerial chase between the pair that has been called Sexual-chase. This is characterized by the female's dashing off with elusive flight and the male's following close behind. The male often gives the Chase-call. The chase may be extended or short; it may stay in the territory or range over several territories; and it may end with both birds' just stopping or with the male's actually pecking the female with his beak. In some cases the chase attracts other males in adjoining territories, and it becomes a group chase of the one female. Territorial borders are momentarily ignored, but once the chase is over, all males quickly return to their own areas.

The third interaction between mates starts with the male doing Song-flight over the territory and then diving down in among the reeds. While landing, he holds his wings in a V above his head and gives a harsh, rasping, drawn-out call like *chjjjjjj*. As in so many cases of behavior-watching, the call will draw your attention to the rest of the display. Following landing, the male may walk or waddle among the reeds with his wings still stiffly spread in a V and possibly continue the call. In some cases the female flies to him and follows him through the vegetation.

During mating, both the male and the female may give the Teetee-call. The female crouches with head and tail raised and wing-quivers. The male approaches with body feathers ruffled and tail spread and lowered; as he gets near he may raise his wings and flutter them slightly. He steps on the back of the female and coition takes place.

Male Redwings are polygamous, averaging three mates per breeding season. The mates of a given male are rarely at the same stage of their breeding cycles; rather, it is more common that one female arrives and starts nesting, and then another female is accepted onto the territory and starts her breeding cycle slightly later. The pair bond does not last past the breeding season; after breeding Redwings remain in separate-sex flocks.

Nest-Building

Placement: In shrubs or among reeds and grasses, often suspended from a number of vertical supports, 3–8 feet high
Size: Inside diameter about 3 inches
Materials: Woven with grasses and lined with finer grasses

The nest is built entirely by the female. She collects grasses and reeds from the area and weaves them around vertical supports so that the nest is actually suspended. It is usually well camouflaged, but it is made obvious by the behavior of the female and her mate, who both call loudly whenever you approach it.

Locating the Nest

WHERE TO LOOK In cattail marshes, at the edges of water where there is lush vegetation, or among the weeds of meadows

WHEN TO LOOK From late April through July

BEHAVIORAL CLUES TO NEST LOCATION:

1. Watch the territorial behavior of a male, for the nest is in his territory.
2. Watch where females are flushed from dense vegetation, for they fly straight up from the nest when disturbed.
3. Watch for the male and female flying above you and giving Check-call or Tseert-call. This means you are near the nest.

Breeding

Eggs: 3–5
Incubation: 11 days, all by the female
Nestling phase: 11 days
Fledgling phase: 7–10 days
Broods: 1

Egg-Laying and Incubation

Before the period of egg-laying, both the male and the female often leave the territory to feed elsewhere; but once most of the eggs have been laid and the female starts to incubate, both birds remain almost exclusively on the territory. The female does all the incubation, and the young hatch about eleven days after the laying of the last egg.

For the rest of the breeding cycle, both the male and female are quick to scold if you come near the nest. The female flies off the nest and gives the Ch'ch'chee-chee-call, while the male gives the Check-call and Tseert-call. Both continually Tail-flick throughout the time of danger. Crows may come into the marsh to eat eggs or young, and in these cases males often leave their territories and band together to chase out the intruders.

Nestling Phase

The female does the majority of caring for the young, bringing food to the nest, and carrying away fecal sacs. If the female is returning to the nest with food and she sees a possible predator (often it may be you), she will remain five to ten yards from the nest and Wing-flip with either one or both wings. If the male approaches, the female does even more vigorous Wing-flipping and stops only when he leaves. When the disturbance is gone, she continues on to the nest. The young remain in the nest for about eleven days.

Fledgling Phase

When ten or eleven days old, the young crawl out of the nest and perch in cover nearby. They are then fed by both parents, the male often taking a greater part in the feeding than he did at earlier stages. The young stay near the nest for a week or more and may remain nearby even after the parents have stopped feeding them. When they leave, they form flocks with other juvenile Redwings and feed and roost together.

Plumage

Redwings go through one complete molt per year, from late July through early September, and during this time all Redwings seem to disappear. Actually they are still present, but they recede into the marshes, keeping out of sight among the reeds throughout the day. When the molt is complete the birds emerge once again.

The different plumage of the male and female reflects the different roles that they play during the breeding season. The female is cryptically colored and spends most of her time among the reeds building the nest and raising the young. The male sits above the reeds and spends the majority of his time defending the territory with displays that utilize his bright red epaulets. The male is able to expose or cover up his epaulets. Thus he can use the bright color in displays but is not made more conspicuous by it in times of danger.

The sexes can always be distinguished by their plumage, the male being black and the female streaked brown.

Seasonal Movement

Red-Winged Blackbird males are some of the first birds to migrate north in spring. They arrive on the breeding ground a few weeks ahead of the females. The females arrive either with or slightly ahead of the first-year males and females.

The next group movement of Redwings takes place in August during the molt, when the birds leave the breeding grounds for more secluded areas. Just before the fall migration the males reappear on the marshes and then move south in large flocks, sometimes joined by Grackles and Cowbirds. The females and immature birds follow in a few weeks.

Common Grackle / *Quiscalus quiscula*

COMMON GRACKLES ARE CLOSELY RELATED TO RED-WINGED BLACK-birds, and as you observe the behavior of the two species, notice the similarities. They both do Bill-tilt and a form of Sexual-chase; and the Ruff-out display of the Grackle is very similar to the Songspread of the Redwing. But other aspects of their lives are extremely different. The Redwing aggressively defends its territories, but the Grackle tends to live in colonies and defends only a small area around the nest. Grackles are harder to observe, since they do not restrict their movements to within a territory.

Therefore, as you start out watching Common Grackles, do not try to follow the life of a particular pair, but watch the general progress of a whole nesting group. In this way you will see most of the birds' common displays and will see more examples of the different stages of their breeding cycle.

Small groups of Grackles making short flights and then landing together are typical of the early stages of courtship. They usually involve one female and a number of males. The males can be distinguished from the females in these flights because they use the V-tail-flight display, whereas the females do not. When the female lands, the males all land near her and often begin to display toward each other, using the Bill-tilt and Ruff-out displays. As pair formation continues, the number of males following a particular female decreases until there is just one male to one female.

Early in the season you are likely to see Grackles flying with long strands of grass in their beaks. The birds may do this up to a month before the nest is really built. The final nest is quickly done by the female, and her active nest-building is a good clue that the various stages of breeding are about to occur. Grackle nests are often too inaccessible to observe closely, but by knowing the length of each phase of breeding you can be aware of when to look for behavioral changes in the adults and when to look for the fledglings leaving the nest.

BEHAVIOR CALENDAR

	TERRITORY	COURTSHIP	NEST-BUILDING	BREEDING	PLUMAGE (MOLTS)	SEASONAL MOVEMENT	SOCIAL BEHAVIOR
JANUARY							
FEBRUARY							
MARCH							
APRIL							
MAY							
JUNE							
JULY							
AUGUST							
SEPTEMBER							
OCTOBER							
NOVEMBER							
DECEMBER							

DISPLAY GUIDE

Visual Displays

Ruff-Out

Male or Female *Sp Su*

With tail and wings slightly spread and contour feathers ruffled, the bird rises up by extending its legs and gives a variable squeaking call.

CALL Song

CONTEXT Given by paired birds to each other, and given by males to each other when group flights land or during territorial skirmishes; may also be given by males during group flights. *See* Courtship

Bill-Tilt

Male or Female *Sp Su*

The bill is tilted up above the horizontal plane, sometimes as a quick toss but more often as a posture that is held for a number of seconds.

CONTEXT Done most often by males when one bird approaches another too closely; has the effect of making one bird eventually move. *See* Courtship

Head-Down

Male *Sp*

The bird's head is lowered and its bill is pointed to the ground. The wings and tail are spread and the contour feathers ruffled.

CALL Song alternated with *peepeepeepeep*

CONTEXT Occurs prior to mating. *See* Courtship

V-Tail-Flight

Male *Sp*

The male folds his tail vertically like the keel of a boat and holds it that way. This is most prominent at the beginnings and ends of flights. From the side, the bird's tail looks like a wedge rather than like the normal thin line.

CONTEXT Done by males in breeding condition, especially in group flights; may also occur while bird is walking. *See* Courtship

Wing-Quiver

Female *Sp*

The female crouches down, draws her head in, and tilts her bill up, while her wings quiver close to her sides.

CALL Seee-call

CONTEXT Given early in the season when group flights land, after conflicts with other birds, and prior to and during mating. *See* Courtship

Auditory Displays

Song

Male or Female *Sp Su F W*

reedeleek or *scoodeleek* or *ch'gasqueek*

An extremely variable, raspy and harsh call, often ending with a high squeak; best identified by the accompanying Ruff-out display

CONTEXT Each bird has its own distinctive version, which it keeps throughout the season. The call is given with the Ruff-out display, and is voiced by females mostly in pair interactions and by males in aggressive or pair interactions. *See* Courtship

Chack-Call

Male or Female *Sp Su F W*

A short, harsh call much like its written approximation; accompanied by a flick of the tail and wings. Males may have a slightly lower version of the call than females. *chaaack*

CONTEXT Given in many diverse situations; especially prevalent during flights to and from the colony

Chaa- and Chitip-Call

Mostly Female *Sp Su F W*

Chaa-call: slightly drawn out and diminishing to a close; Chitip-call: a two-part phrase of two harsh notes *chaaa* and *chitip*

CONTEXT Most often associated with taking flight or aggressive encounters; both calls used in similar contexts

Brrt-Call

Male *Sp Su F W*

An extremely short call with a sharp beginning and ending *brrrt*

CONTEXT Occurs in a variety of circumstances, often causing other birds to give the same call

Seee-Call

Male and Female *Sp*

A drawn-out, high-pitched whistle; not loud *seeee*

CONTEXT Given by male and female during mating. The female gives it during Wing-quiver, and the male may alternate it with the Song during the Head-down display. *See* Courtship

BEHAVIOR DESCRIPTIONS

Territory

Type: Nest site area
Size: 4–8 square yards
Main behavior: Perch-taking
Duration of defense: From before nest-building to fledgling phase

The only aspect of the Common Grackle's behavior that could be called territorial occurs in the immediate area of the nest. Once two birds have paired, they begin looking for a nest site. After the site is chosen a few grasses are brought there, and it is defended primarily by the female. But she is only occasionally near the site, since she feeds and mates elsewhere. Generally, other Grackles that land within a yard of the nest are chased off, or more usually, the nest owner merely flies at the spot where the intruder is and takes its perch. The male perches near the female and may help her defend the nest site if another pair of Grackles comes too close.

Many nests are often built within the same few trees, forming nesting colonies. This behavior obviously precludes the use of large territories.

Courtship

Main behavior: Singing groups, group flights, mutual singing, and Head-down
Duration: 1–6 weeks

When Common Grackles first arrive on the breeding ground, they may spend part of each morning in "singing groups." These are composed of as many as twenty birds, all perched in the bare branches of a tree and all singing at the same time. The Songs can be quite loud and are accompanied by the Ruff-out display in both the female and the male. Any one singing group may last for only a half hour, and during that time its membership changes as birds come and go.

Two more common aspects of Grackle courtship are "group flights" and "mutual singing." Group flights are started by a female's leaving her perch and being followed in the air by two or more males. The flights are roughly of two types: "leader flights," where a number of males fly slowly behind and to the side of the female, and "chase flights," where the female flies quickly and elusively and is followed just as quickly by the males. During the early part of the breeding season, the males can be easily distinguished from the females in these flights, for they hold their tails in a vertical V — a display known as V-tail-flight, which may advertise sexual maturity. Later in the season and outside the breeding season, the males do not fly with V-tail-flight, and it is harder to distinguish between the sexes.

When group flights land, the males often display to each other with Bill-tilt or Ruff-out. These probably function as competitive displays and may indicate that in group flights, males are competing among each other for the female. During male displays the female may do Wing-quiver, a display more commonly associated with mating.

As the breeding season continues, group flights diminish in size until there is just a pair of birds, one male and one female. This is usually an indication that these birds have paired and will breed. The male and female spend much of their time together from now until incubation and often engage in mutual singing, where each alternates Song with the other.

Mating involves striking displays that will be seen about the time the nest is complete. If it is initiated by the female with Wing-quiver, the male may be stimulated to do Head-down and to orient himself alongside the female. He then steps onto her back and they copulate. If the male initiates the Head-down display, the female will often attack him, walk away, or ignore him. Mating displays may occur on the ground or in trees.

An aspect of Grackle pairing, unusual among our common

birds, is that a significant proportion of males desert their mates during the period of incubation and pair with other females. The period of pair formation with these new females is generally shortened. The first female continues to raise her brood but does all the feeding herself.

Nest-Building

Placement: On horizontal branches (often in evergreens) 3–30 feet high
Size: Inside diameter 4 inches
Materials: Grasses with a middle layer of mud

Grackle nest-building may occur over a period of up to six weeks or even more, and in these extended cases seems to be closely tied with pair formation. Grackles are somewhat colonial in their nesting habits, many pairs often nesting in the same area. After a pair have become established they spend most of their time at a breeding site, first just exploring: visiting old nests and hopping about prospective nest sites. During these activities the female is always in the lead.

Their behavior changes when they have chosen a nest site, for then they restrict most of their activities to the area of the site, at least during the times they are at the breeding area.

When the owners are present, they defend this site against other Grackles that land nearby. (*See* Territory.) During this stage both male and female will be seen carrying long strands of nesting material and draping them over their chosen nest location. This phase of the nesting cycle is the most variable, lasting from one to four weeks with no real building going on.

A sign that a new phase has started is increased nest-building activity by the female. She will start building in earnest and will be followed by the male during her trips to gather material. The bulk of the nest will be finished in about five days and lined with fine grasses. The male does not help build in this final phase of construction.

Locating the Nest

WHERE TO LOOK Mostly in groves of evergreens in the city or the country, usually near open areas and often in the same place from year to year

WHEN TO LOOK During mid-spring, especially in April

BEHAVIORAL CLUES TO NEST LOCATION During the first stage of partial building, both male and female will be seen carrying conspicuously long grasses to the nest site. Follow them and you will probably have located a breeding site where many other Grackles will also be nesting. The first stage of building is not always where the final nest will be, and the nests can be hard to see among the green foliage, especially if they are in spruces.

Breeding

Eggs: About 5
Incubation: 12 days average, done by the female only
Nestling phase: 12 days average
Fledgling phase: Possibly none
Broods: 1

Egg-Laying and Incubation

About three to four days after the completion of the nest, the female begins laying eggs. She lays one each day until the clutch is complete, usually at about four to five eggs. Incubation is done solely by the female and lasts for eleven to thirteen days. During this time the male may guard the nest as the female leaves to feed. Also during this time, the male may leave the female and start a shortened pair formation with a second female. In such cases he rarely returns, and the lone female then raises the brood by herself.

Nestling Phase

If the male stays with his original mate, then both participate in feeding the young when they hatch. Fecal sacs are removed by both parents and dropped away from the nest or eaten. The nestling phase lasts about twelve days.

Fledgling Phase
Young Grackles can fly well soon after leaving the nest and are fed by the parents for only a few days. After that, they join with other juveniles and form large flocks that feed and roost together.

Plumage

Common Grackles go through one complete molt per year, starting in August and September. The coloration of the bird does not change throughout the year.

The male tends to have more iridescence on his head and neck region than the female, and this may be to accentuate his more pronounced Ruff-out display and Head-down display, where these feathers are fluffed out. The slightly longer tail in the male is used in the V-tail-flight display, the main behavioral distinction between the sexes during the breeding season. At other times of year it is hard to tell male from female.

Seasonal Movement

Grackles are extending their range to the north and to the west continually. In general, they are migratory, moving north mainly in March and April and returning south mainly in September and October. In winter some individuals or small flocks remain farther north than most of their species.

Other large movements of Common Grackles involve the flights of huge flocks to and from the roosting sites. These occur every morning and evening, and, depending on the area, may involve anywhere from twenty or thirty birds to thousands of birds. (*See* Social Behavior.)

Social Behavior

Grackles are known for their large communal roosts. In the more southern areas of the birds' range, these roosts may be permanent areas that are used throughout the year. In more northern areas, they are used primarily from the middle of breeding season until fall migration, and again in early spring.

The roosts often have thousands of birds in them, and as the birds arrive for the night they may make spectacular formations in flight, one of the most common being a long line of birds trailing off into the distance. The Grackles leave the roosts in large flocks during the day and feed together in fields and along roadsides.

During the breeding season the roosts diminish in size and may contain only nonbreeding birds or mated males whose partners are on the nest incubating during the night. As the first broods hatch and become independent, they join the roosts, and in turn are joined by later broods and by females that are finished with nesting. The roosts are largest just before migration and may contain other species such as Starlings, Redwings, Cowbirds and Robins.

Feeder Behavior

Common Grackles seem to frequent feeders primarily during spring and fall migration, but they may also appear at other times. They feed on a variety of seeds, but they especially like cracked corn scattered on the ground.

Most common displays: Bill-tilt is very commonly seen, and is given by both males and females. The bird doing the most vertical tilt of its bill is usually the most dominant bird. Chaa-call and Chitip-call are also heard during aggressive encounters. See the Display Guide.

Other behavior: In spring you may see many features of courtship near or at your feeders. See the section on Courtship.

American Goldfinch / *Spinus tristis*

THROUGHOUT WINTER, AMERICAN GOLDFINCHES REMAIN IN SMALL flocks and are generally quiet except for their *perchicoree* call during flight. But in spring, at about the same time that the male grows his bright yellow plumage, their behavior changes. The males begin to sing a warbling song from prominent perches, and there are numerous chases between males, and between males and females. This short period is an interesting time in the life cycle of the Goldfinches, for they do not start nest-building until late summer, and for the time in between, they are relatively quiet.

The beginning of their nest-building and breeding is marked by the presence of two unusual flight displays by the male. Both of these flight patterns differ markedly from the normal shallow, looping flight of all Goldfinches. One is a deep looping flight like the pattern of a roller coaster, and the other is a flat stalling flight with an explosive warbling call. Both of these flights are exciting to see, and after you are familiar with them they can become, along with the calls of cicadas and the blooms of goldenrod, a real sign of the heart of summer.

The female is secretive about building the nest, but you must try to find it so that you can enjoy seeing the relationship of the pair during incubation. The male approaches the nest with one of the flight displays and circles overhead. If the female calls, the male drops down and feeds her near the nest. This is an exciting behavior to see, but you may have to wait for it to occur, for the intervals between feedings can be as long as an hour.

It is easy to tell when the first Goldfinch young have fledged, for they follow the parents about in the air and have a distinctive call that sounds exactly like the written words *chipee, chipee,* which is very different from the normal adult flight call of *perchicoree, perchicoree.* For me, just as the Goldfinch singing marks spring and the flight displays mark midsummer, so do the calls of the Goldfinch fledglings mark the end of the breeding season and the imminence of fall colors and cooler weather.

BEHAVIOR CALENDAR

	TERRITORY	COURTSHIP	NEST-BUILDING	BREEDING	PLUMAGE (MOLTS)	SEASONAL MOVEMENT	SOCIAL BEHAVIOR
JANUARY							
FEBRUARY							
MARCH							
APRIL							
MAY							
JUNE							
JULY							
AUGUST							
SEPTEMBER							
OCTOBER							
NOVEMBER							
DECEMBER							

DISPLAY GUIDE

Visual Displays

Flat-Flight

Male *Sp Su*

The male flies in a flat slow flight with deep fluttering wingbeats. Flights last for five to ten seconds and are often fifty feet or more above the ground. The accompanying call is distinctive.

CALL Long-song

CONTEXT Given in spring and early summer, in open areas and in the presence of another male or female Goldfinch; sometimes given after skirmishes between males. *See* Courtship

Deep-Loop-Flight

Male *Su*

An exaggeration of the normal looping flight of all Goldfinches. It alternates deep dives, where the wings are held in, with coastings upward, where the wings are held out.

CALL Flight-call

CONTEXT Generally given by a male before and during the breeding period; given sometimes as the male circles about the boundary of his territory. *See* Territory, Breeding

Chase-Flight

Male(s) and Female *Sp Su*

One or more males fly after a single female. The female may be flying in an elusive, zigzagging pattern.

CONTEXT Done in spring, during supposed period of pair formation; also in summer on a male's territory, in which case the birds speed in and out of the trees and shrubbery. *See* Courtship

Auditory Displays

Long-Song

Male *Sp*

A warbling, reminiscent of a canary's song; lasts up to thirty seconds or more and is given repeatedly with short pauses in between.

CONTEXT Given by males from exposed perches, especially during midday in late spring. *See* Courtship

Short-Song

Male *Sp Su*

A short warbling song seemingly sputtered out; lasts only two to three seconds and then is repeated after a five-to-eight-second pause.

CONTEXT Given in spring near other males and females, and in summer from perches around the edge of the territory. *See* Courtship, Territory

Flight-Call

Male and Female *Sp Su F W*

A light phrase with three or four syllables;

most easily identified by the fact that it accompanies almost all adult flying

perchicoree. perchicoree. perchicoree.

CONTEXT Almost always heard as the birds fly; occurs on the upswing of their normal looping flight, although during nesting the females may fly silently

Sweeet-Call

Male or Female *Sp Su F W*

A short, ascending whistle, given softly, but easily heard from quite a distance

sweeet or *sweeyeet*

CONTEXT Given in situations where there is the possibility of danger or when there is a minor conflict with another Goldfinch; heard most often near the nest and at the winter feeder. *See* Nest-building

Bearbee-Call

Male or Female *Su F*

A short, two- or three-part phrase very much like its written description; ranges from barely audible to easily heard; usually repeated a number of times

bearbee. bearbee, bee. bearbee.

CONTEXT Given especially near the nest when there is extreme danger such as an aerial or ground predator. *See* Nest-building

Feeding-Call

Female *Su F*

A high, delicate call composed of a rapid series of short whistles; hard to hear unless you are close

teeteeteeteetee

CONTEXT Given by the female while she is on the nest and the male is flying overhead; often induces the male to fly down to the nest and feed the female. *See* Breeding

Fledgling-Call

Male and Female — *Su F*

chipee. chipee. chipee pee.

A short, two-part call, the second part having a definite *ee* sound, the first sound being short and harsh; accent on the second part CONTEXT Given by fledglings during flight as they follow one or both parents; may also be heard just after they alight; not heard in late fall after young are independent of their parents. *See* Breeding

BEHAVIOR DESCRIPTIONS

Territory

Type: Nesting
Size: 1/4 acre to an acre or more
Main behavior: Short-song, circular flights, chases
Duration of defense: From a week before nest-building to when the nest is completed

The territorial behavior of male Goldfinches takes three forms: Short-song given from a few perches around a limited area; the chasing of other males off this area; and high display flights roughly circling over the area.

The circular flights are conspicuous and are a good clue to territory locations. As a bird circles high in the air it usually does one of the flight displays — Deep-loop-flight is the most common, but Flat-flight is also regularly seen. In some cases Short-song is the most conspicuous aspect of territoriality, being given frequently throughout the day from three or four perches. When another male Goldfinch perches in the area, the territorial male often stops singing, and both birds sit silently, watching each other. Then the territorial male flies at the intruder and takes his perch. If the intruder does not leave immediately he may be chased. Both the Short-song and the chases are also seen in spring, but at this time they are not associated with a particular area and are more likely a part of courtship.

Territorial behavior in Goldfinches is most prominent in crowded areas where the territories may be as small as one hundred feet in diameter. Generally, males defend their territories against other males, while females defend the immediate area of the nest site against other females. When incubation starts, territorial boundaries begin to dissolve, and by the time of the nestling phase, other Goldfinches can perch and feed near the nest and not be bothered.

Courtship

Main behavior: Long-song, Short-song, chases, Flat-flight
Duration: Variable, from spring into early summer

Some radical changes in Goldfinch behavior occur in spring and are believed to be associated with courtship. The new auditory displays are the most obvious and will draw your attention to other interesting behavior.

All winter the birds have given only two calls: their Flight-call and the Sweeet-call. Now the two types of warbling songs are heard. Both are given from exposed perches by a male, and generally other males or females will be nearby. Often a number of males are near one female, and at some point she flies off and all the males chase after her. There may also be fights between males, where both face each other and flutter together high into the air. After these skirmishes, one of the males may then do Flat-flight in the immediate area. The fights between males are often short, but the chases of the female by one or more males may occur over a large area and continue for twenty minutes or more.

These patterns of behavior are most common in spring, long before the Goldfinches start nesting, but they will also be seen occasionally throughout the summer. During summer, Goldfinches become less conspicuous as they fly about and feed together in small groups. This behavior continues until territory formation and nest-building begin.

Mating takes place later in summer, around the time of nest-building and usually near the area of the nest. No unusual displays are associated with it.

Nest-Building

Placement: In upright forks of leafy bushes or trees; about 4–20 feet high
Size: Inside diameter 2 inches
Materials: Bark strips from weeds and vines, filaments from wind-dispersed seeds, webbing from caterpillars or spiders

As the female Goldfinch collects materials and builds the nest, she remains secretive, flying silently rather than with the normal Flight-call and staying within the cover of vegetation. Thistle flowers and the web nests of caterpillars are two common places where the female gathers materials, and following her as she leaves these sites may help you to locate the nest. The male either flies along with her on collecting trips or else remains perched near the nest. Once the nest is completed, both the male and the female leave the area. At first it seems as if they have abandoned the nest and their territory, but they return in a few days, and the female starts to lay eggs.

During nest-building and again during breeding, two calls are given by Goldfinches when there is danger near the nest: the Sweeet-call, given during minor disturbances, and the Bearbee-call, usually given in cases of greater danger. Sometimes these calls are given in response to your presence and will be a signal that you are near the nest.

Locating the Nest

WHERE TO LOOK In open areas, either marshy or near water, with scattered shrubs and/or saplings

; down to near the nest and the female joins him
ge quantities of partially digested seeds. The male
d the female returns to the nest.

e

st week the nestlings are quiet in the nest and
when the parent arrives, but will silently reach
en the edge of the nest is tapped. In the second
active about the nest and call loudly when the
At this time the young may change their excre-
d instead of defecating in the nest and letting the
way the fecal sacs, the young back up to the edge
defecate just over its rim. Because of this habit,
often have a coating of droppings on their outer
before the young fledge they develop a soft *chick-*
call that becomes louder once they leave the nest.
both parents during the nestling phase, receiving
ds from the parents' crops. Feedings are at long
here from thirty to ninety minutes apart.

e

e nest, the young are fed primarily by the male,
male has already started a nest for her second
glings typically give the Fledgling-call and this
locate them for feeding. They also give the
they follow their parents in flight. Thus, family
distinguished from flocks made up just of adults,
the adult Flight-call will be heard. As soon as
e totally independent of their parents they no
call.

Plumage

hrough two molts per year. There is a complete
ers in fall after breeding, and then there is a
pring of all but the wing and tail feathers. The

WHEN TO LOOK From about
most active nest-building is
the last week in August

BEHAVIORAL CLUES TO NEST I

1. Follow female if she h
just leaving webbed caterpil

2. Watch the female's r
nonfeeding area, i.e., in an
seeds. This usually indicate

3. Look for territorial bel
flights and then dropping to
male in a nonfeeding area.

Eggs: 5
Incubation: 12–14 days, by t
Nestling phase: 11–15 days
Fledgling phase: Variable, pc
Broods: 1–2

Egg-Laying and Incuba

Eggs are generally laid a
but in some cases the bi
About five eggs are laid,
the laying of the second
to fourteen days.

If you see the female
then you know she is m
study found the female
of her time on the nest
tion is matched by few c
she can do this is that th
to feed her. This is a m
often using one of the f
about once every hour.
lovely light call that so

the male drop
and accepts la
then leaves a

Nestling Pha

During the fi
make no nois
up for food w
week they ar
parent arrives.
tory habits, ar
parents carry
of the nest an
Goldfinch nest
layer. The day
kee or *che-wee*
They are fed b
regurgitated se
intervals, anyw

Fledgling Pha

When out of th
for often the f
brood. The fle
helps the male
Fledgling-call a
flocks are easily
from which on
the fledglings a
longer give this

Goldfinches go
molt of all feat
partial molt in

molts produce little change in the colors of the female, but they alter the male's appearance noticeably. In winter he looks like the female, but in summer he is easily distinguished, for he is a brighter yellow and has a black crown. When you see the male's plumage change in spring, start looking for the behavioral changes associated with courtship.

In spring and summer the sexes can be distinguished by behavior as well as by plumage, for the male is the only one to do the display flights and the two types of Song.

Seasonal Movement

Like many other common birds, Goldfinches vary in their degree of seasonal movement. They are believed to be year-round residents in most of North America, but there are also marked movements of Goldfinch flocks in spring and fall. These flocks may contain from ten to several hundred birds.

Social Behavior

Goldfinches commonly feed in flocks during fall and winter. When they arrive at feeders you may see some aggressive postures exchanged between birds that are feeding close together. The main one is a threat posture with head forward and the bill gaping. Such postures may help establish the dominance of some birds over others without actual fighting taking place. The only calls heard are the Flight-call and occasionally the Sweeet-call.

Feeder Behavior

Goldfinches are attracted to thistle and hulled sunflower seeds. They come to feeders all year. In early spring, you can see courtship behavior, and in late summer the parents may bring the young to the feeder. See the sections on Courtship and Breeding (Fledgling Phase).

FL

Song Sparrow / *Melospiza melodia*

THE EASIEST ASPECT OF SONG SPARROW BEHAVIOR TO APPRECIATE is the territorial behavior of the male. In areas where the birds migrate, the male arrives on the breeding ground ahead of the female and starts to define a territory by singing his rich, warbling song from three or four prominent perches. Once you locate a territory you can always count on the bird being there, for he does all of his feeding as well as nesting and mating in that area.

As the season progresses and more males arrive, look for territorial skirmishes — new males challenging established ones for favored areas. These take place with an unusual display, where one of the birds puffs out his feathers, possibly raises one or both wings, and sings. The challenges to a territory may continue for an hour or more.

It is important to locate territories and gain a sense of their dimensions while the male is actively advertising them with Song, for once the female arrives the male does much less singing, and the birds are very secretive as they move about the territory together. Even locating the birds at this time can be difficult, but if you have learned the extent of the territory earlier you will know where to look for their activity.

Following Song Sparrows in anything other than their territorial behavior takes real patience. This is especially true during the stages of breeding when the adults are feeding the young. They become cautious about approaching the nest, and you will have to move quite far away before they even tentatively go about their normal routine. The nest is made of grasses and may be right on the ground, so it can be hard to see, even when you are looking directly at it.

BEHAVIOR CALENDAR

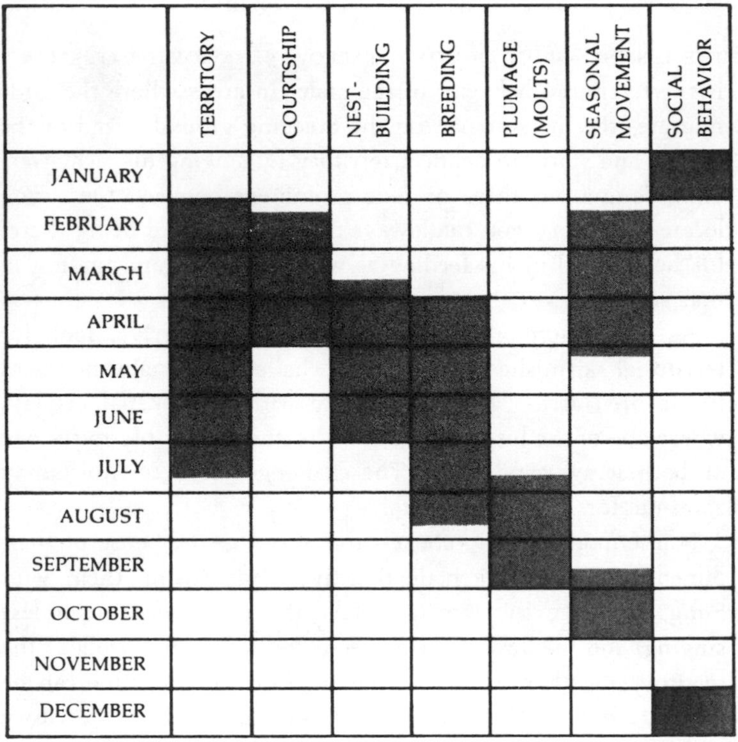

	TERRITORY	COURTSHIP	NEST-BUILDING	BREEDING	PLUMAGE (MOLTS)	SEASONAL MOVEMENT	SOCIAL BEHAVIOR
JANUARY							
FEBRUARY							
MARCH							
APRIL							
MAY							
JUNE							
JULY							
AUGUST							
SEPTEMBER							
OCTOBER							
NOVEMBER							
DECEMBER							

DISPLAY GUIDE

Visual Displays

Puffed-Out

Male *Sp Su*

The bird's contour feathers are fluffed, and he takes a horizontal posture.

CONTEXT Given during territorial disputes by either a territory holder or an intruder, depending on their dominance status at that moment; generally given by the defender first; also given during neighbor interactions along a common border. *See* Territory

Puff-Sing-Wave

Male *Sp Su*

The contour feathers are fluffed, and one or both wings are raised and possibly vibrated.

CALL Song

CONTEXT Occurs during territorial challenges. *See* Territory

Pouncing

Male and Female *Sp Su*

The male suddenly swoops down onto a female and sings a loud song as he flies by. If the female is his mate, she sits still and gives the Trill-call. A foreign female will react to Pouncing with fighting and threat calls.

CALL Song, Trill-call, Zhee-call

CONTEXT Occurs between a male and his mate and between a male and his neighbors' mates before egg-laying starts, and just before, in the case of the second brood. *See* Courtship

Crest-Raise

Male or Female *Sp Su F W*

The crest feathers are fluffed and the wings and/or tail may be flicked.

CALL Tchunk-call

CONTEXT Occurs in situations of possible danger

Auditory Displays

Song

Male *Sp Su F*

A rich and varied warble starting with a few repeated notes; variable in individuals

CONTEXT Given by males from exposed perches in their territories. A soft version is given during Puff-sing-wave; a loud version is given to call the female off the nest; and a version with a twittering beginning can be given when the bird is in flight from one perch to another. *See* Territory, Breeding

Tsip-Call

Male or Female *Sp Su F W*

tsip. tsip.

A short, soft call

CONTEXT Given between mates as they move about together, possibly helping them to stay in contact. *See* Courtship

Tchunk-Call

Male or Female *Sp Su F*

tchunk. tchunk.

A deeper and louder call than Tsip-call

CONTEXT Given during times when there is possible danger in the area; self-assertive

Zhee-Call

Male or Female *Sp Su F*

A growllike call *zheeee. zheeee.*

CONTEXT Used as a close-distance threat against another Song Sparrow; warns of possible attack

Trill-Call

Female *Sp Su*

A high-pitched, thin trill

CONTEXT Given by female when Pounced by her mate, when greeting a male, and following copulation. *See* Courtship

BEHAVIOR DESCRIPTIONS

Territory

Type: Mating, nesting, and feeding
Size: ½–1½ acres
Main behavior: Singing from perches, Puffed-out, and Puff-sing-wave
Duration of defense: From late winter to the onset of molting in August

The start of male Song in spring is the best sign of the beginning of territory formation. The bird will spend long periods of each day singing from three or four perches around its territory. If you look closely you will see that the brown spot on the breast of the bird is ruffled out as it sings. The male will at first claim a much larger area than it will finally defend, as is the case with many other bird species.

Juvenile males in spring are allowed to hop around within an adult's territory and feed, as long as they don't give territorial displays. If they act territorial they will be quickly chased away. These juveniles have a light warbling song at this stage, but when they decide to claim a territory this song quickly changes to the adult form.

Transients that are migrating through are immediately flown at if they try to land in a male's established territory, and they usually leave before any contact takes place. Neighboring males will be continually calling in response to the territory holder, and the two males may meet near borders, where one or both may do Puffed-out, sometimes followed by a short chase.

A foreign male that intrudes on a territory with the purpose of taking over part of it is met with the Puff-sing-wave display. The territory holder moves along close to the ground, his body feathers fluffed, singing softly and possibly lowering and raising one wing. The intruder follows him silently. After a while a short chase or fight may occur, after which both birds move to safe areas and sing loudly. This sequence may be repeated until the intruder either is routed or is able to claim some of the owner's territory. Often both birds will appear to be feeding throughout the interaction.

Territorial defense wanes near the end of the second brood and ceases during late summer when molting takes place. In fall there is renewed territorial activity, especially on the warmer days — males will be heard singing, and there will be some territorial disputes. Most males will then migrate south, but a small proportion will stay in the area of their territories throughout the winter. After fall there is no Song or defense of the territories until late winter or spring.

Courtship

Main behavior: Song, Pouncing, Trill-call, decrease in Song
Duration: Approximately 2 weeks

When looking for a new phase of behavior you must first know what the previous activities of the bird were in order to realize what has changed. This is particularly true when the new behavior pattern involves only subtle changes. The pair formation of Song Sparrows is not marked by conspicuous displays, and in fact the female is quite secretive when she arrives on the territory. When the male first discovers a female on his territory he treats her as he would any potential territorial intruder; he may fly at her and even scold. But she responds differently from other intruders, and this distinguishes her as a female. As the male dives at her, she does not flee but stays still and gives the Trill-call. In time the male seems to accept her presence, and the two birds move about the territory together and follow their daily routines.

Generally at this stage the male greatly reduces his early-morning singing, to only about ten Songs per hour. To notice this difference you must listen to the male's singing early in the season before the female arrives. This decrease can be the clue to the female's arrival, since she is often too secretive to locate.

Once nest-building has started, other behavior is more frequent. This includes renewed singing by the male, copulation attempts, and Pouncing. Pouncing is the only behavior of Song Sparrows that resembles a pair-formation display. Here the male dives down upon the female and may even hit her; then he sings loudly and continues on in flight.

The female gives the Trill-call at the male as he dives, especially early in the season; later on she gives the scolding Zhee-call. Pouncing takes place only up until egg-laying time, then it stops. If it starts again slightly later, then most likely the first nesting was unsuccessful. Pouncing will occur again just before the egg-laying of the second brood.

The male will also pounce on neighboring females, but they will respond with fighting and scolding calls, which in turn attract their mates to defend them.

As the pair move about the territory you may hear the slight Tsip-call, which seems to keep the pair in aural contact while they may be visually isolated.

Nest-Building

Placement: From ground level up to about 4 feet, among shrubs or high weeds
Size: Inside diameter 2¼ inches
Materials: Mostly grasses, some bark strands, and leaves

There are two clues to the start of nest-building: one is to see either the male or the female with nesting material in its beak; and the other is to catch glimpses of new behavior patterns. These include: copulation attempts, Pouncing, and renewed singing by the male, which increases until he is singing almost as much as he was during territory formation. These may be the only indications you will have that the nest is being built, for the female often stays well within cover as she collects material and builds. Although the male may carry nesting material about, the female does all building of the final nest. About two days are needed to complete the nest's main structure, and then another few days to add the lining of softer grasses, fine rootlets, or hair.

It is common for the birds to have to start nest-building a number of times, for during the early stages of incubation or egg-laying, the nests are frequently discovered by predators.

Locating the Nest

WHERE TO LOOK In open areas where there is both low vegetation and some shrub growth; old lots, old fields, or at the edges of swamps, lakes or rivers

WHEN TO LOOK Spring and midsummer, for there is usually more than one brood

BEHAVIORAL CLUES TO NEST LOCATION:

1. Get an idea of the extent of the male's territory and look in that area.
2. Look for the female building when the male resumes Song after being quieter during pair formation.
3. The birds generally scold you as you approach the nest area. Simply move far away and watch them with binoculars, and they

will repeatedly go to the nest. They do not enter the nest directly, but enter by way of dense cover a yard or two from it.

Breeding

Eggs: 3–5
Incubation: 12–13 days, by the female only
Nestling phase: 10 days
Fledgling phase: 20 days
Broods: 2–3

Egg-Laying and Incubation

When the nest is finished, the eggs are laid, one each day, until the clutch is complete. The female does all the incubation. All Pouncing by the male stops once egg-laying has started. During the incubation period the female has a rhythm of leaving the nest about once every half hour for five to ten minutes and then returning. While the female is incubating, the male never approaches closer than a few yards from the nest. A little more than half of the times that the female leaves the nest during incubation, it seems to be in response to actions of the male. The male interrupts his normal singing, flies to a perch a few yards from the nest, and gives several loud versions of his Song. Often the female leaves the nest at this point and feeds. Other times she may make no visible response to the male or may remain on the nest and give the Trill-call.

Nestling Phase

The nestling phase is slightly shorter for Song Sparrows than it is for other common birds of comparable size. The young are brooded for the first half of the nestling phase by the female, and both male and female help in feeding them. The young Song Sparrows leave the nest before they can fly well.

Fledgling Phase

When the young first leave the nest they move only short distances from it and remain there quietly for about the first week, being fed regularly by the parents. During this time, the young are hard to find.

Studies have shown that in some populations the parents share the labor of feeding the young in such a way that the brood is roughly divided between the parents, each feeding its own set of fledglings.

After this, when the young are able to fly, they are easily spotted as they follow after the parents, wing-fluttering and begging with loud notes. During the end of this phase, the young may become more antagonistic toward each other and chase and fight. They can be easily distinguished from the adults, for their tails are only half grown at this stage.

The first egg of the second brood may be laid as soon as six days after the young of the first brood have started their fledgling phase. In this case, the male does most of the feeding for the fledglings of the first brood.

Plumage

Song Sparrows have one complete molt per year, usually occurring from mid-August to the end of September for the males, and slightly later for the females. Territorial behavior in the male tapers off just before the molt begins and then may start again at a reduced level when the molt is completed. This change can be followed by listening for the presence or absence of Song, which in the Song Sparrow is directly linked with territorial behavior.

The plumage of male and female is identical. Song is the main characteristic for distinguishing the sexes.

Seasonal Movement

The majority of Song Sparrows are migratory, taking long flights to the south in fall and back north in spring. Spring migration has been observed to take place in roughly two waves: the first birds to arrive on the breeding ground are the males that have bred there in the past; the second wave consists of females and first-year males. This spring migration takes place from mid-February to mid-March.

Not all Song Sparrows migrate south during winter; many remain on the breeding ground throughout the year. Of these, most are adult males that remain in the areas of their breeding territories. In the middle latitudes of North America, up to fifty percent of the adult males may be residents. Only about twenty percent of the adult females remain through the winter in the same areas.

Some males that remain as residents one year migrate the next year, and some migrants later become residents. Thus the individual Song Sparrow's migratory habits are not fixed.

Glossary
Bibliography

Glossary

Bill-wipe — Bill is wiped across a branch during confrontations
Brood — The birds hatched from one clutch of eggs
Brooding — The act of sitting over the newly hatched young to keep them warm; done by the parents in the first few days of the nestling phase
Call — An auditory display, generally simpler in structure than song
Courtship — All behavior that involves the relationship between males and females in breeding condition
Display — A stereotyped movement or sound that a bird makes, and that, when used in certain situations, affects the behavior of other animals about the displaying bird
Fecal sac — A small mass of excrement, surrounded by a coating of mucus, which is excreted by a nestling and carried off or eaten by the parents
Fledgling — A young bird that has left the nest but is still dependent on its parents for some or all of its food
Home range — An area inhabited by a bird, but not necessarily defended against its own or other species
Incubation — The act of covering the eggs to keep them warm and further their development
Mate-feeding — The feeding of one adult member of a pair by the other; usually occurs only during the breeding season (sometimes called "courtship feeding")
Nestling — A hatched bird that remains in the nest and is cared for by the parents or other adults
Pair formation — The aspect of courtship that involves the pair's first encounters and their becoming committed to each other
Primary roost — A fixed location at which birds habitually gather during the inactive phase of their day
Range — An area regularly inhabited by a bird, but not consistently defended
Seasonal movement — Predictable large-scale movement of populations over the course of the year
Secondary roost — Like the primary roost, but used for a shorter period of time and during the active phase of the birds' day
Song — A complex auditory display that may be partially inherited and partially learned
Territory — Any defended area

Bibliography

This bibliography includes only the major works that were used in the research for this guide. Other works can be located by looking through the bibliographies of those books mentioned here. The amount of study that has been done on each species varies widely and is reflected generally in the number and length of the studies listed for each bird.

Canada Goose

Blurton Jones, N. G. 1960. Experiments on the causation of the threat postures of Canada Geese. *Wildfowl Trust Ann. Rept.* 11: 46–52.

Collias, N. E., and Jahn, L. R. 1959. Social behavior and breeding success in Canada Geese (*Branta canadensis*) confined under semi-natural conditions. *Auk* 76: 478–509.

Johnsgard, P. A. 1965. *Handbook of waterfowl behavior.* Ithaca, N.Y.: Cornell Univ. Press.

Klopman, R. B. 1962. Sexual behavior in the Canada Goose. *Living Bird* 1: 123–29.

———. 1968. The agonistic behavior of the Canada Goose. *Behavior* 30: 287–319.

Radesater, T. 1974. Form and sequential associations between the triumph ceremony and other behavior patterns in the Canada Goose *Branta canadensis* L. *Ornis Scandinavica* 5: 87–101.

Raveling, D. G. 1970. Dominance relationships and agonistic behavior of Canada Geese in winter. *Behavior* 37: 291–319.

Mallard

Abraham, R. L. 1974. Vocalizations of the Mallard. *Condor* 76: 401–20.

Collias, N. E. 1962. The behavior of Ducks. In *The behavior of domestic animals,* ed. E. S. E. Hafez. Baltimore: Williams and Wilkins Co.

Hochbaum, H. A. 1944. *The Canvasback on a prairie marsh.* Wash., D.C.: Amer. Wildl. Inst.

Johnsgard, P. A. 1960. A quantitative study of sexual behavior of Mallards and Black Ducks. *Wilson Bull.* 72: 133–55.

———. 1965. *Handbook of waterfowl behavior.* Ithaca, N.Y.: Cornell Univ. Press.

Lebret, T. 1958. Inciting ("Hetzen") by flying Ducks. *Ardea* 46: 68–72.
———. 1961. Pair formation in the annual cycle of the Mallard, *Anas platyrhynchos* L. *Ardea* 49: 7–157.
Lorenz, K. 1953. Comparative studies on the behavior of the *Anatinae*. *Avicult. Mag.* 59: 1–87.
———. 1958. The evolution of behavior. *Sci. Am.* (Dec.) 199: 67–78.
McKinney, F. 1965. The comfort movements of *Anatinae*. *Behavior* 25: 120–220.
———. 1969. The behavior of Ducks. In *The behavior of domestic animals,* ed. E. S. E. Hafez. Baltimore: Williams and Wilkins Co.
Raitasuo, L. 1964. Social behavior of the Mallard, *Anas platyrhynchos,* in the course of the annual cycle. *Paper on Game Res.* (Helsinki) 24: 1–72.
Titman, R. D., and Lowther, J. K. 1975. The breeding behavior of a crowded population of Mallards. *Can. J. Zool.* 53: 1270–83.
Simmons, K. E. L., and Weidmann, U. 1973. Directional bias as a component of social behavior with special reference to the Mallard, *Anas platyrhynchos. J. Zool.* (London) 170: 49–62.
Weidmann, U., and Dailey, J. 1971. The role of the female in the social display of Mallards. *Anim. Behav.* 19: 287–98.

American Kestrel

Balgooyen, T. G. 1976. Behavior and ecology of the American Kestrel *(Falco sparverius)* in the Sierra Nevada of California. *Univ. of Calif. Publ. Zool.* 103: 1–83.
Cade, T. J. 1955. Experiments on the winter territoriality of the American Kestrel (*Falco sparverius*). *Wilson Bull.* 67: 5–17.
Collopy, M. W. 1977. Food caching by female American Kestrels in winter. *Condor* 79: 63–68.
Mills, G. S. 1976. American Kestrel sex ratios and habitat separation. *Auk* 93: 740–48.
Mueller, H. C. 1971. Displays and vocalizations of the Sparrow Hawk. *Wilson Bull.* 83: 249–54.
Willoughby, E. G., and Cade, T. J. 1964. Breeding behavior of the American Kestrel. *Living Bird* 3: 75–96.

Herring Gull

Bent, A. C. 1921. Life histories of North American Gulls and Terns. *U.S. Natl. Mus. Bull.* 113: 102–19.
Moynihan, M. 1956. Notes on the behavior of some North American Gulls. *Behavior* 10: 126–78.
———. 1958*a*. Notes on the behavior of some North American Gulls. Ibid. 12: 95–182.
———. 1958*b*. Notes on the behavior of some North American Gulls. Ibid. 13: 113–30.
Paludan, K. 1951. Contributions to the breeding biology of *Larus argentatus* and *Larus fuscus. Vidensk. Medd. Dansk Naturh. Foren.* 114: 1–128.
Tinbergen, N. 1952. On the significance of territory in the Herring Gull. *Ibis* 94: 158–59.
———. 1953. *The Herring Gull's world.* London: Collins.

Pigeon

Fabricius, E., and Jansson, A. 1963. Reproductive behavior of the Pigeon (*Columbia livia*). *Anim. Behav.* 11: 534–47.

Goodwin, D. 1955. Notes on European wild Pigeons. *Avicult. Mag.* 61: 54–85.

———. 1967. *Pigeons and Doves of the world.* London: British Museum.

Mathews, L. H. 1939. Visual stimulation and ovulation in Pigeons. *Proc. Roy. Soc.* 126: 557–60.

Chimney Swift

Bent, A. C. 1940. Life histories of North American Cuckoos, Goatsuckers, Hummingbirds and their allies. *U.S. Natl. Mus. Bull.* 176: 271–93.

Fisher, R. B. 1958. The breeding biology of the Chimney Swift. *N.Y. State Mus. Bull.* 368: 1–141.

Pickens, A. L. 1935. Evening drill of Chimney Swifts during the late summer. *Auk* 52: 149–53.

Common Flicker

Kilham, L. 1959. Early reproductive behavior of Flickers. *Wilson Bull.* 71: 323–36.

Lawrence, L. De K. 1967. A comparative life-history study of four species of Woodpeckers. *Amer. Ornith. Union, Ornith. Monogr.* 5: 1–156.

Noble, G. K. 1936. Courtship and sexual selection of the Flicker (*Colaptes auratus iuteus*). *Auk* 53: 269–82.

Sherman, A. R. 1910. At the sign of the Northern Flicker. *Wilson Bull.* 17: 135–66.

Hairy Woodpecker

Jackson, J. A. 1976. How to determine the status of a Woodpecker nest. *Living Bird* 15: 205–21.

Kilham, L. 1966. Reproductive behavior of Hairy Woodpeckers. Part 1. Pair formation and courtship. *Wilson Bull.* 80: 286–305.

———. 1969. Reproductive behavior of Hairy Woodpeckers. Part 3. Agonistic behavior in relation to courtship and territory. Ibid. 81: 169-83.

Lawrence, L. De K. 1967. A comparative life-history study of four species of Woodpeckers. *Amer. Ornith. Union, Ornith. Monogr.* 5: 1–156.

Eastern Kingbird

Davis, D. E. 1941. The belligerency of the Kingbird. *Wilson Bull.* 53: 157–68.

Hausman, L. A. 1925. On the utterances of the Kingbird. *Tyrannus tyrannus* L., with especial reference to a recently recorded song. *Auk* 42: 220–26.

Smith, W. J. 1966. Communication and relationships in the genus *Tyrannus*. *Publ. Nuttall Ornith. Club* no. 6: 1–67.

Tree Swallow

Bent, A. C. 1942. Life histories of North American Flycatchers, Larks, Swallows, and their allies. *U.S. Natl. Mus. Bull.* 179: 11–28.

Kuerzi, R. G. 1941. Life history studies of the Tree Swallow. *Proc. Linnaean Soc. New York* 52–53: 1–52.

Schaeffer, F. S. 1970. Observation of "billing" in the courtship behavior of the Tree Swallow. *Bird Banding* 41: 242.

Stocek, R. F. 1970. Observations on the breeding biology of the Tree Swallow. *Cassinia* 52: 3–20.

Blue Jay

Bent, A. C. 1946. Life histories of North American Crows, Jays, and Titmice. *U.S. Natl. Mus. Bull.* 191: 32-52.

Hardy, J. W. 1961. Studies in behavior and phylogeny of certain New World Jays. *Univ. of Kansas Science Bull.* 42: 13-149.

Laskey, A. R. 1958. Blue Jays at Nashville, Tennessee — movements, nesting, age. *Bird Banding* 29: 211-18.

Common Crow

Bent, A. C. 1946. Life histories of North American Crows, Jays, and Titmice. *U.S. Natl. Mus. Bull.* 191: 226-59.

Chamberlain, D. R., and Cornwell, G. W. 1971. Selected vocalizations of the Common Crow. *Auk* 88: 613-34.

Good, E. E. 1952. The life history of the American Crow — *Corvus brachyrhynchos brehm.* Ph.D. dissertation, Ohio State Univ.

Goodwin, D. 1976. *Crows of the world.* Ithaca, N.Y.: Cornell Univ. Press.

Black-Capped Chickadee

Dixon, K. L., and Stefanski, R. A. 1970. An appraisal of the song of the Black-Capped Chickadee. *Wilson Bull.* 82: 53-62.

Ficken, M., and Witkin, S. R. 1977. Responses of Black-Capped Chickadee flocks to predators. *Auk* 95: 156-57.

Ficken, M.; Ficken, R. W.; and Witkin, S. R. 1978. Vocal repertoire of the Black-Capped Chickadee. *Auk* 95: 34-48.

Glase, J. C. 1973. Ecology of social organization in the Black-Capped Chickadee. *Living Bird* 12: 235-67.

Holleback, M. 1974. Behavioral interactions and the dispersal of the family in Black-Capped Chickadees. *Wilson Bull.* 86: 466-68.

Odum, E. P. 1941*a*. Annual cycle of the Black-Capped Chickadee. *Auk* 58: 314-33, 518-35.

———. 1941*b*. Annual cycle of the Black-Capped Chickadee. Ibid. 59: 499-531.

Smith, S. T. 1972. Communication and other social behavior in *Parus carolinensis. Publ. Nuttall Ornithol. Club* no. 11.

Stefanski, R. A. 1967. Utilization of the breeding territory in Black-Capped Chickadees. *Condor* 69: 259-67.

House Wren

Kendeigh, S. C. 1941. Territorial and mating behavior of the House Wren. *Illinois Biol. Monogr.* 18: 1-120.
———. 1952. Parental care and its evolution. *Illinois Biol. Monogr.* 22: 1-93.

Mockingbird

Hailman, J. P. 1960. Hostile dancing and fall territory of a color-banded Mockingbird. *Condor* 62: 464-68.
———. 1963. The Mockingbird's "Tail-up" display to mammals near the nest. *Wilson Bull.* 75: 414-17.
Hicks, T. W. 1955. Mockingbird attacking Blacksnake. *Auk* 72: 296-97.
Horwich, R. H. 1969. Behavioral ontogeny of the Mockingbird. *Wilson Bull.* 81: 87-93.
Laskey, A. R. 1933. A territory and mating study of Mockingbirds. *Migrant* 4: 29-35.
———. 1935. Mockingbird life history studies. *Auk* 52: 370-81.
———. 1936. Fall and winter behavior of Mockingbirds. *Wilson Bull.* 43: 241-55.
———. 1962. Breeding biology of Mockingbirds. *Auk* 79: 596-606.
Michener, H., and Michener, J. R. 1935. Mockingbirds, their territories and individualities. *Condor* 37: 97-140.
Michener, J. R. 1951. Territorial behavior and age composition in a population of Mockingbirds at a feeding station. *Condor* 53: 276-83.

Gray Catbird

Barlow, J. C. 1963. Notes on an epigamic display of the Catbird. *Wilson Bull.* 75: 273.
Bent, A. C. 1948. Life histories of North American Nuthatches, Wrens, Thrashers, and their allies. *U.S. Natl. Mus. Bull.* 195: 320-51.
Berger, A. J. 1954. Injury-feigning by the Catbird. *Wilson Bull.* 66: 61.
Slack, R. D. 1976. Nest guarding behavior by male Gray Catbirds. *Auk* 93: 292-300.
Zimmerman, J. L. 1963. A nesting study of the Catbird in southern Michigan. *Jack-Pine Warbler* 41: 142-60.

American Robin

Howell, J. C. 1942. Notes on the nesting habits of the American Robin (*Turdus migratorius* L.). *Am. Mid. Nat.* 28: 529-603.
Young, H. 1951. Territorial behavior in the eastern Robin. *Proc. Linn. Soc. N.Y.*, nos. 58-62: 1-37.
———. 1955. Breeding and nesting of the Robin. *Am. Mid. Nat.* 53: 329-52.

Starling

Brodie, J. 1976. The flight behavior of Starlings at a winter roost. *Brit. Birds* 69: 51-60.

Bullough, W. S. 1942. The reproductive cycles of the British and continental races of the Starling. *Phil. Trans. Roy. Soc. London* Ser. B 231: 165-246.

Davis, E. D. 1959. Territorial rank in Starlings. *Anim. Behav.* 7: 214-21.

Ellis, C. R., Jr. 1966. Agonistic behavior in the male Starling. *Wilson Bull.* 78: 208-24.

Hamilton, W. J., III, and Gilbert, W. J. 1969. Starling dispersal. *Ecology* 50: 886-98.

Hartby, E. 1965. The calls of the Starling (*Sturnus vulgaris*). *Dansk Orn. Foren. Tidsskr.* 62: 205-30.

Jumber, J. R. 1956. Roosting behavior of the Starling in central Pennsylvania. *Auk* 73: 411-26.

Kessel, B. 1957. A study of the breeding biology of the European Starling (*Sturnus vulgaris* L.) in North America. *Am. Mid. Nat.* 58: 257-331.

Ward, P., and Zahavi, A. 1973. The importance of certain assemblages of birds as "information centres" for food-finding. *Ibis* 115: 517-34.

Red-Eyed Vireo

Barlow, J. C., and Rice, J. C. 1977. Aspects of the comparative behavior of Red-Eyed and Philadelphia Vireos. *Can. J. Zool.* 55: 528-42.

Lawrence, L. De K. 1953. Nesting life and behavior of the Red-Eyed Vireo. *Can. Field-Nat.* 67: 47-87.

Southern, W. E. 1958. Nesting of the Red-Eyed Vireo in the Douglas Lake region, Michigan. *Jack-Pine Warbler* 36: 103-30, 185-207.

Common Yellowthroat

Hofslund, P. B. 1959. A life history study of the Yellowthroat, *Geothlypis trichas*. *Proc. Minn. Acad. Sci.* 27: 144-74.

Stewart, R. E. 1953. A life history study of the Yellow-Throat. *Wilson Bull.* 65: 99-115.

House Sparrow

North, C. A. 1973. Movement patterns of the House Sparrow in Oklahoma. *Ornithol. Monogr.* 14: 79-91.

Sappington, J N. 1977. Breeding biology of House Sparrows in north Mississippi. *Wilson Bull.* 89: 300-9.

Simmons, K. E. L. 1954. Further notes on House Sparrow behavior. *Ibis* 97: 478-81.

Summers-Smith, D. 1954. The communal display of the House Sparrow *Passer domesticus*. *Ibis* 96: 116-28.

———. 1955. Display of the House Sparrow *Passer domesticus*. *Ibis* 97: 296-305.

———. 1963. *The House Sparrow*. London: Collins.

Weaver, R. L. 1939. Winter observations and a study of the nesting of English Sparrows. *Bird-Banding* 10: 73-79.

———. 1945. Reproduction in English Sparrows. *Auk* 60: 62-74.

Red-Winged Blackbird

Blakley, N. R. 1976. Successive polygyny in upland nesting Redwinged Blackbirds. *Condor* 78: 129–33.

Nero, R. W. 1956. A behavior study of the Red-Winged Blackbird. *Wilson Bull.* 68: 4–37, 129–50.

Orians, G. H., and Cristman, G. M. 1968. A comparative study of the behavior of Red-Winged, Tricolored, and Yellow-Headed Blackbirds. *Univ. of Calif. Publ. in Zool.* 84.

Peek, F. W. 1972. An experimental study of the territorial function of vocal and visual display in the male Red-Winged Blackbird (*Agelaius phoeniceus*). *Anim. Behav.* 20: 112–18.

Common Grackle

Ficken, R. W. 1963. Courtship and agonistic behavior of the Common Grackle, *Quiscalus quiscula*. *Auk* 80: 52–72.

Maxwell, G. R., II, and Putnam, S. P. 1972. Incubation, care of young, and nest success of the Common Grackle (*Quiscalus quiscula*) in northern Ohio. *Auk* 89: 349–59.

Peterson, A., and Young, H. 1950. A nesting study of the Bronzed Grackle. *Auk* 67: 466–76.

Wiens, J. A. 1965. Behavioral interactions of Red-Winged Blackbirds and Common Grackles on a common breeding ground. *Auk* 82: 356–74.

Wiley, R. H. 1976. Affiliation between the sexes in Common Grackles. *Z. Tierpsychol.* 40: 59–79, 244–64.

———. 1976. Communication and spatial relationships in a colony of Common Grackles. *Anim. Behav.* 24: 570–84.

American Goldfinch

Coutlee, E. L. 1967. Agonistic behavior in the American Goldfinch. *Wilson Bull.* 79: 89–109.

Drum, M. 1939. Territorial studies on the Eastern Goldfinch. *Wilson Bull.* 51: 69–77.

Mousley, H. 1930*a*. The home life of the American Goldfinch. *Can. Field-Nat.* 44: 177–179, 204–7.

———. 1930*b*. The home life of the American Goldfinch. Ibid. 46: 200–3.

Mundinger, P. C. 1972. Annual testicular cycle and bill color change in the Eastern American Goldfinch. *Auk* 89: 403–19.

Nice, M. M. 1939. "Territorial Song" and non-territorial behavior of Goldfinches in Ohio. *Wilson Bull.* 51: 123.

Stokes, A. W. 1950. Breeding behavior of the Goldfinch. *Wilson Bull.* 62: 106–27.

Walkinshaw, L. H. 1938*a*. Life history studies of the Goldfinch. *Jack-Pine Warbler* 16: 3–11, 14–15.

———. 1938*b*. Life history studies of the Goldfinch. Ibid. 17: 3–12.

Song Sparrow

Knapton, R. W., and Crebs, J. R. 1976. Dominance hierarchies in winter Song Sparrows. *Condor* 78: 567–69.

Nice, M. M. 1937. Studies in the life history of the Song Sparrow. *Trans. Linn. Soc. N.Y.*, 4.

———. 1941. Studies in the life history of the Song Sparrow. Ibid. 6.

Smith, J. N. M. 1978. Division of labour by Song Sparrows feeding fledged young. *Can. J. Zool.* 56: 187–91.

Tompa, F. S. 1962. Territorial behavior: the main controlling factor of a local Song Sparrow population. *Auk* 79: 687–97.

Brief Index